FOURTH
EDITION

Insights

A LABORATORY MANUAL FOR

Physical Geology

Clair Russell Ossian *Tarrant County College*

KendallHunt
publishing company

Book Team

Chairman and Chief Executive Officer   Mark C. Falb
President and Chief Operating Officer   Chad M. Chandlee
Vice President, Higher Education   David L. Tart
Senior Managing Editor   Ray Wood
Assistant Editor   Bob Largent
Editorial Manager   Georgia Botsford
Developmental Editor   Melissa M. Tittle
Vice President, Operations   Timothy J. Beitzel
Assistant Vice President, Production Services   Christine E. O'Brien
Senior Production Editor   Carrie Maro
Senior Permissions Editor   Colleen Zelinsky
Cover Designer   Suzanne Millius

Cover image © 2010 PhotoDisc.

Kendall Hunt
publishing company

www.kendallhunt.com
Send all inquiries to:
4050 Westmark Drive
Dubuque, IA 52004-1840

BRIEF CONTENTS

CONTENTS

ACKNOWLEDGMENTS

The completion of this laboratory manual was in part due to the generosity of individuals and companies in making many copyrighted images and text extracts available for illustrations. The text portions are identified where they occur in the manual but thanks also goes to the following publishing companies: Addison, Wesley, Longman (UK); American Geological Institute; Cambridge University Press; Canfield Publishing; Harvard University Press; John Wiley and Sons; McGraw-Hill Companies, Inc.; MIT Press; Rand McNally & Company; Random House; Simon and Schuster, Inc. (Free Press); University of Chicago Press; and W. H. Freeman and Company.

Everyone is touched by individuals who have inspired them and taught them important parts of their skills. Among these special professors and friends have been the late Carl Gugler (University of Nebraska), a zoologist who taught me to look for the unusual aspects of my trade and gave me my first sense of the richness of life, and Sam Trevis (University of Nebraska), who taught me how to use systematic scientific approaches as he gave me the tools to understand mineralogy and petrology. J. Alan Holman (Michigan State University) thought he saw something special and gave me that first great chance to start down the road to becoming a scientist; Alan Scott (University of Texas–Austin) assisted me in comprehending the many implications of sedimentary depositional environments; and, finally, Wann Langston (University of Texas–Austin), who honed my understanding of the scientific process, was a source of encouragement as well as a good friend.

I have further benefited from conversation and interaction with my fellow instructors at Tarrant County College, David Kotila and Hayden Chasteen, both of whom have patiently contributed to the editing of this manual.

Lastly, I offer a most sincere "thank you" to my wife Eleanor who through the years has always believed in me, always understood me, always encouraged me, and who has given up so much of our life together so that I might travel, do fieldwork, and learn the many skills needed to become a successful geoscientist. I can't imagine having taken this journey without her.

This manual was prepared to fill a need for a broad, general approach to core curriculum geology topics in support of physical geology.

Some instructors will wish to cover more types of topics, such as fluvial or coastal systems, glaciology, geochemistry, and others, but most curricula focus on a few fundamental topics that we all understand are essential for both beginning geology students and for those students who joined this class because they had a general interest in the subject.

Topics covered in this book will prepare a student for field identification of rocks, minerals and will offer students a strong foundation for the understanding and making of maps. The keys and approaches found in this manual not only teach the student to recognize earth materials, they teach them how to systematically identify them, using a scientific approach, not simply memorizing their general appearance. If we instructors can accomplish these tasks during the brief time we have these students with us, we can feel that we have done our job well.

No geologist can be taken seriously if he or she does not know Steno's Laws, has not mastered the basis for Uniformitarianism, cannot understand the nature of the evidence for the age of the earth, or has remained ignorant of Bowen's Reaction series and the huge implications of the rock cycle. No book or laboratory manual can do it all, and this one does not pretend to be comprehensive. It is a general survey of the basics of geology, and as such, it will be a powerful tool for students.

As the author, I hope for constructive critical feedback from instructors and from students. Together we can turn what I believe to be a good manual into a superb one. I will respond positively to all such attempts to refine and improve this work.

Clair Russell Ossian is a Professor of Geology at Tarrant County College, Hurst, Texas. His career includes nearly twenty years in geological research as a paleontologist and sedimentologist, experiencing many challenges world wide while working in the petroleum industry, retiring as Principal Research Geologist from ARCO Oil and Gas Co. Field research during that time involved projects in diverse settings, including Alaska, Canada, China, Indonesia, Norway, Africa and numerous projects within the rest of the United States.

Before, during and after that phase of his career, he has continued to teach Physical Geology, Historical Geology, Environmental Geology and Field Courses. Dr. Ossian holds a Bachelor of Science degree in Geology from the University of Nebraska (Lincoln), a Master of Science degree from Michigan State University, and a Doctor of Philosophy from The University of Texas (Austin). Both graduate degrees are in geology and paleontology.

Professor Ossian has published numerous papers on sedimentology, vertebrate paleontology, invertebrate paleontology and his current research is centered on the geology of Egypt. His most recent research involved providing geological support for an archaeological excavation in the Egyptian delta.

He is a member of the American Association for the Advancement of Science, the Society of Sigma XI, the American Research Center in Egypt (ARCE), the Egyptian Exploration Society and is a Certified Geologist with the American Association of Petroleum Geologists.

A long time resident of north Texas, he and his wife have raised two children, a son and a daughter. They currently share their home with a pair of Chinese chowchow dogs. Hobbies include gardening, raising Japanese koi fish, growing water lilies, carpentry, photography and writing. Travel continues to be a great pleasure for them, proving many opportunities for new experiences . . . and new geological insights.

1

Minerals

■ IDENTIFICATION AND PROPERTIES

Minerals are arguably the most important aspect of your Physical Geology laboratory experience. They are found in every aspect of our lives and make up all rocks. Once you understand how minerals form and how they make up the earth's materials, your textbook readings will become much more meaningful.

Glance around your laboratory to confirm the importance of minerals. The floor, walls, ceiling, lighting . . . and even your clothes are products of natural minerals obtained and refined.

Some minerals are constructed of atoms and molecules that form beautiful gems like diamonds, emeralds, and rubies . . . but much more useful minerals occur as ores of metals, materials used in wallboard, paper, and concrete. More than 3000 minerals are now known to scientists, and the number continues to grow. The majority are obscure and seldom seen, so we will concentrate on those minerals most common and useful to society.

The definition of a mineral is fundamental. Geologists generally agree that a mineral must:

1. **Be naturally occurring.** No materials known only from the laboratory or from industry will qualify. Example: There are a few man-made materials that are believed to be slightly harder than diamonds, but they are manufactured and therefore do not qualify as minerals.

2. **Be inorganic.** Minerals are normally considered NOT to have been the product of some living organism or composed largely of compounds of carbon. Example: Coal made from partially decomposed plant materials cannot be a mineral.

3. **Have a definite chemical composition.** Each mineral has a defined and distinctive chemical composition. Example: NaCl is the chemical formula for ordinary table salt, the mineral halite.

   If we were able to prepare a hypothetical compound whose formula could be written as Na_2Cl (bisodium chloride), it would almost certainly have different properties such as taste, color, crystal shape, etc., and would no longer be halite. (But such a compound is impossible due to the limits of atomic size and chemical bonding.)

4. **Have distinctive and definitive physical properties.** Mineral appearance and character is controlled by the arrangement of atoms and molecules. Example: Cleavage of halite is controlled by the packing order and atomic size of sodium and chlorine atoms. These atoms fit so that halite crystals can only break along planes at right angles to each other, forming **cubes** (see later discussion about cleavage).

5. **Be a crystalline solid or have characteristic internal structures** (atomic arrangement). Minerals could be identified solely from texture, mass, color, and general appearance, but, if you used this method, you would never be sure whether the mineral you discovered on your vacation was one you already knew or one that you had never seen before. It is better to study diagnostic features of minerals and to learn to distinguish them through the use of identification keys.

<table>
<tr><td colspan="4">Which of These Is a Mineral?</td></tr>
<tr><td colspan="4">Using the criteria in the list above, check out these common items. Some of these items qualify but others do not. Why did some of them fail the test?</td></tr>
<tr><td>Table Salt</td><td>___________</td><td>Emerald</td><td>___________</td></tr>
<tr><td>Sugar</td><td>___________</td><td>Plastic</td><td>___________</td></tr>
<tr><td>Coal</td><td>___________</td><td>Copper</td><td>___________</td></tr>
<tr><td>Glass</td><td>___________</td><td>Plaster of Paris</td><td>___________</td></tr>
</table>

This chapter presents a list of characters known to be most useful in identifying minerals and will teach you to quickly and accurately determine their names.

The tests are all simple, because you won't have laboratory balances, hardness points, or other apparatus while standing on the side of that tall mountain with an unknown mineral specimen in your hand.

To acquire these skills, you need the following materials:

- A tray with 25 to 35 mineral specimens
- A dropper bottle containing 3 to 5 percent hydrochloric acid
- A copper penny
- An ordinary wire carpenter's nail
- A hardened masonry nail
- A glass plate
- An unglazed porcelain streak plate
- A magnet
- A pencil
- An eraser

We first need to build a vocabulary of useful terms. Ask your instructor to help apply these concepts as you sort minerals by the major character features described below such as streak, luster, hardness, or specific gravity. When you learn to recognize these features, you will quickly develop the ability to distinguish between even very similar minerals.

Mineral Features Useful for Identification
Crystal Shape

Crystals are naturally occurring structures that grow by the addition of atoms on smooth faces of minerals. These surfaces (faces) that are produced give a regular shape that is often characteristic for a given type of mineral. For instance, quartz crystals are

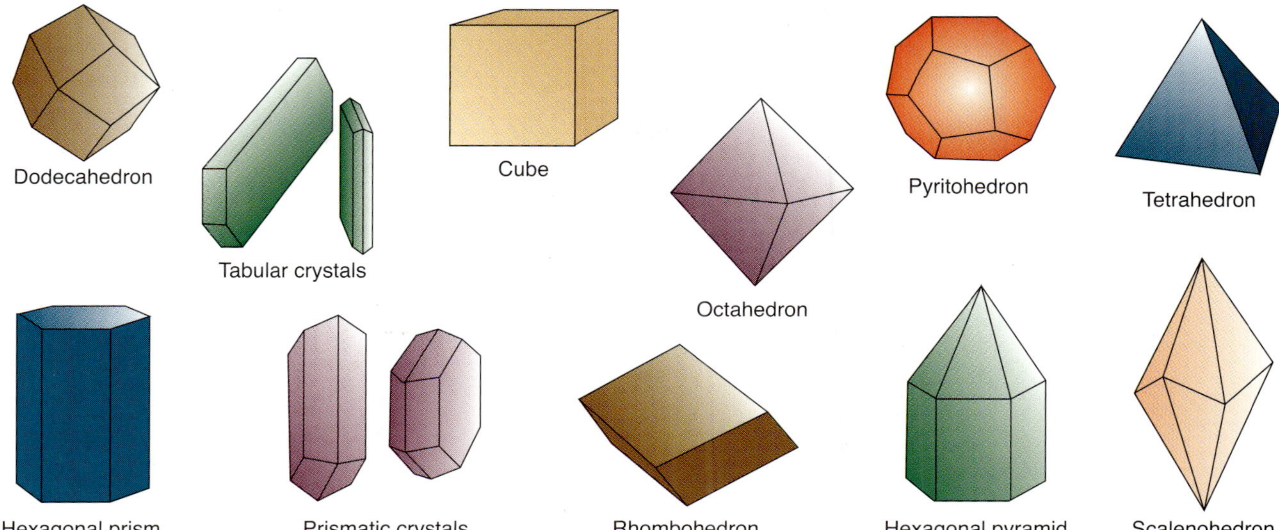

FIGURE 1.1 Common crystal habits.

elongated, six-sided shapes (***hexagonal***), while calcite crystals have a distinctive "leaning" rectangular shape known as a ***rhombohedron*** (see Figure 1.1). One of the most important identification tools a mineralogist can use is the fact that the angles between any two adjacent faces of a particular mineral are always the same.

When crystals meet the precise angular rule above, we say that the crystal is ***euhedral***—that is, extremely regular and uniform in shape. (See Figure 1.3.)

FIGURE 1.2 It is important to remember that the same crystal shape can occur in more than one kind of mineral. In this photograph of the mineral galena, note that the crystal in the upper left corner is an octohedron, while others are cubes (upper right), and still more show combinations in between the two shapes. All of these dark gray crystals have the same chemical formula and are the same mineral.

FIGURE 1.3 Photograph of anhedral (left), subhedral (center), and euhedral crystals (right).

Crystals often lack the open space within a rock that allows them to grow into perfect euhedral shapes. The crystal may have one or several of the sides pressed against other crystals, and those sides are then suppressed or distorted. This is a much more common condition, and these crystals are called **subhedral.**

If mineral grains grow in compact masses so that there is little space for individual crystal faces to develop, then the crystals will lack well-defined euhedral faces and the grain will be irregular in shape. This condition is called **anhedral** and is the most common condition of all.

Fracture and Cleavage

Some minerals have internal atomic order that prevents regular breaking—that is, if you break these minerals, the resulting pieces are irregular. An example would be the fragments produced by dropping a glass bottle onto a hard concrete surface. No two pieces of the broken glass would have exactly the same shape. This manner of breakage is termed **fracture.** Quartz and obsidian are two examples of substances that fracture into odd-shaped pieces (see Figure 1.4), even though crystals of quartz may have beautifully regular shapes.

Cleavage is a very useful feature, because minerals that cleave will break along predetermined planes of weakness generated by the internal atomic arrangement. A piece of the mineral halite has chlorine and sodium atoms arranged to produce three planes of weakness at right angles to each other.

Test this by placing a knife blade against a piece of halite parallel to one of the flat sides and tap the blade with a hammer. Halite breaks smoothly, producing a flat surface. Move the blade in a little toward the center of the halite piece, parallel to the flat surface just produced, and tap it again. The halite will again break, producing a new surface perfectly parallel to the first one.

FIGURE 1.4 Fractured obsidian and cleaved halite.

Cleavages can produce different numbers of cleavage patterns, and cleavage planes often have different angular relationships. Figure 1.5 shows the common types you will work with in this laboratory session.

Micas (biotite and muscovite) are examples of minerals with a single perfect cleavage, a type often called **basal cleavage,** causing them to split into thin parallel sheets. In working with basal cleavage, think of a stack of paper sheets. Each sheet is parallel to the others, and each will lift smoothly, but the only way to divide the sheets across will result in an irregular torn edge. This type of cleavage can also be referred to as "basal pinacoid."

Other minerals cleave into two sets of planes (four parallel faces) with the other two sides of the mineral rough and irregular. Some minerals you test will have two planes at right angles, while others will have a pair of planes at some angle clearly NOT at right angles.

A few minerals have three cleavage planes (six parallel faces) and cleave into cubes, rectangles, or parallelograms called **rhombohedrons. Octahedral** minerals with four cleavage planes (eight parallel faces) are more rare, while those with more than four are seldom seen among the common minerals. The number and arrangement of cleavage planes is one of the best identification clues. (See Figure 1.5.)

It is important to note that some minerals, such as clays, DO TECHNICALLY HAVE cleavage, but their mineral grains and crystals are so small that cleavages can only be seen with electron microscopes.

Identification keys later in the chapter will help guide you to the right answer in these unusual cases.

One way to determine whether cleavage is present is to hold the specimen in your hands and rotate it with a beam of light reflecting from the surface. A bright evident

INSIGHTS

If a mineral displays cleavage, each broken surface will be parallel to those surfaces already present. These parallel broken surfaces are called cleavage planes. Therefore, cleavage planes on opposite sides of the mineral sample should be parallel to each other, or nearly so. If the mineral is a crystal, this may not be true. Place the mineral on the table so it rests on one of the flat surfaces. Is there another flat surface parallel to the table on the top surface of the mineral? If so, you may have found a pair of cleavages that represent one of the cleavage planes. If not, you may be working with a subhedral crystal.

Be aware that crystals may well have paired faces, too, but, in the important examples, the broken ends of the crystal will generally show fractured surfaces.

(NOTE: Two parallel cleaved surfaces in a mineral specimen represent ONE plane of cleavage. Example: A cleaved halite cube has six surfaces representing three pairs of parallel surfaces and therefore has three planes of cleavage.)

Number of Cleavage Directions	Shapes that Crystal Breaks Into	Sketch	Illustration of Cleavage Directions
0 No cleavage, only fracture	Irregular masses with no flat surfaces		None
1	"Books" that split apart along flat sheets		
2 at 90°	Elongated form with rectangular cross sections (prisms) and parts of such forms		
2 not at 90°	Elongated form with parallelogram cross sections (prisms) and parts of such forms		
3 at 90°	Shapes made of cubes and parts of cubes		
3 not at 90°	Shapes made of rhombohedrons and parts of rhombohedrons		
4	Shapes made of octahedrons and parts of octahedrons		
6	Shapes made of dodecahedrons and parts of dodecahedrons		

FIGURE 1.5 Common cleavage patterns in minerals. From *Laboratory Manual in Physical Geology,* 3rd edition by Busch and Tasa, MacMillan Publishers 1996.

flash of reflected light will disclose a smooth surface, often a cleavage surface. Then, rotate the specimen carefully through 180° (rotate it so that it turns over and the opposite side is upwards), and if you get the bright reflection repeated, you are probably looking at the same cleavage plane.

Surfaces will reflect light differently if they are cleaved than they will if they are fractured. Some minerals will cleave with broad smooth surfaces (Figure 1.6) while others may split so that there are many parallel mini-faces that are at different heights. But in this second case, all the parallel steps will still reflect together (Figure 1.7).

Fractured faces will scatter light and never return good clean reflections (Figure 1.8). (Caution: Some materials like quartz have smooth crystal faces that may resemble cleavage surfaces, but broken surfaces are always irregular.)

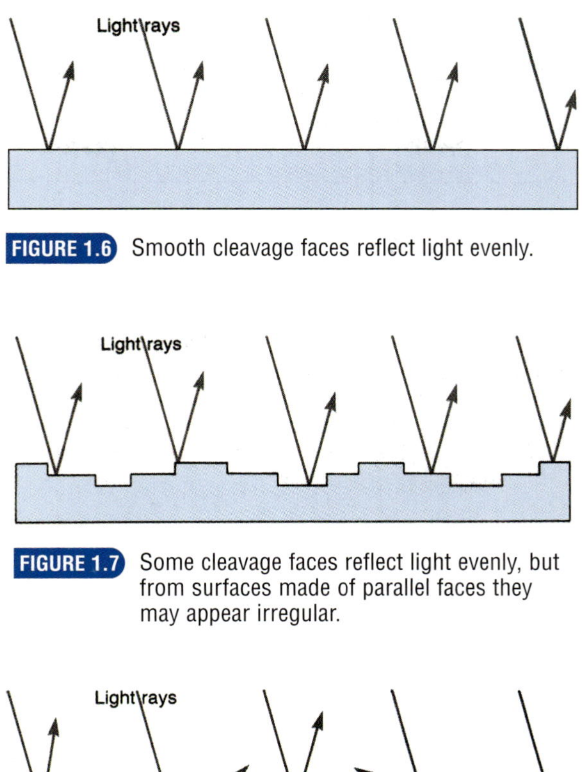

FIGURE 1.6 Smooth cleavage faces reflect light evenly.

FIGURE 1.7 Some cleavage faces reflect light evenly, but from surfaces made of parallel faces they may appear irregular.

FIGURE 1.8 Fractured faces of minerals reflect light unevenly.

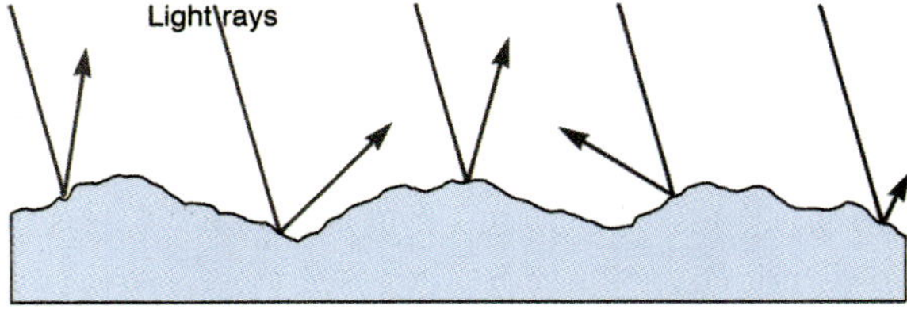

Mohs Scale of Relative Hardness		Useful Aids	
10	Diamond		
9	Corundum		
8	Topaz		
7	Quartz		
6	Orthoclase		
		5.5	Glass, Masonry Nail
5	Apatite		
		4.5	Carpenter's Nail
4	Fluorite		
		3.0–3.5	Copper Penny
3	Calcite		
		2.5	Fingernail
2	Gypsum		
1	Talc		

FIGURE 1.9 Mohs hardness scale.

Note that in Figure 1.6, parallel light rays arrive at a smooth plane surface and the reflected rays all leave at a common angle. This type of cleavage is found in many minerals and is essentially the same effect as that seen when light is reflected off a sheet of glass.

INSIGHTS

SUMMARY: Minerals that display cleavage break into regular shapes, while those that exhibit fracture break into irregular shapes.

INSIGHTS

Hardness and strength are NOT the same thing. Some very hard minerals are fragile and can be easily crushed or broken. (Diamond is such a mineral.)

Because some minerals are brittle, it is often useful to run the hardness test both ways. Press a carpenter's nail into a piece of granular olivine, and because olivine is so brittle and fragile, it will show a clear scratch, even though olivine is hardness 8 and the nail is 4.5. Take olivine and rub it against the nail, and you will see bright shiny steel. The olivine easily scored the metal, demonstrating how much harder it was than the nail. Granular olivine crystals are poorly bonded and come apart when scratched, falsely suggesting that the nail is harder.

Remember, minerals with higher Mohs numbers always scratch those with lower numbers.

Presence or absence of cleavage is one of the most useful characteristics in minerals. Which samples:

SHOW Cleavage		LACK Cleavage
Sample Number	How Many Cleavage Faces?	Sample Number
_______	_______	_______
_______	_______	_______
_______	_______	_______
_______	_______	_______
_______	_______	_______
_______	_______	_______
_______	_______	_______

In Figure 1.7, parallel light rays arrive at a surface, but the smooth plane surface is broken up into many flat surfaces at different heights, but which are still parallel to each other. The arriving light rays will STILL be reflected at a common angle. This type of reflection effect is often seen in minerals such as hornblende.

In Figure 1.8, parallel arriving light rays strike an irregular surface. The reflected rays must then leave the surface in many different directions. This is called 'scattered' light, or disordered light, and is the condition seen in minerals that lack cleavage.

Hardness of minerals is also important and is easily determined thanks to the Mohs Hardness Scale (Figure 1.9). This scale compares the hardness of all known minerals against each other.

Mohs Hardness Scale uses ten common or well-known minerals as memory aids, starting with the softest of all ordinary minerals, talc. A few items commonly at hand help us with the Mohs Hardness Scale. Your fingernails are hardness 2.5; a REAL copper penny is about 3.5 (modern composite pennies can be softer); a standard carpenter's nail is 4.5; and a hardened masonry nail is 5.5. (Be careful, many masonry nails today are much softer than you might have expected.) Another 5.5 hardness tool can be a piece of common glass.

Relative Hardness

Using your sample set and the hardness tools provided, find samples that fit the following criteria. Write the sample number in the blank space:

Hardness of 1	_______	Which mineral is softest?	_______
Hardness of 3	_______	Which mineral is hardest?	_______
Hardness of 5	_______		
Hardness of 7	_______		

Luster is a measure of the way minerals reflect light. Reflection styles include: metallic, nonmetallic, vitreous (glassy), earthy, waxy, glossy, and greasy. Practice determining these surface features with the minerals in your kit.

Specific gravity determines how heavy (dense) a mineral is compared to the density of water. Water has a specific gravity of 1.0, and when we learn that a mineral has a specific gravity of 2.5, we know that, volume for volume, the mineral is 2.5 times as heavy as water.

Determining specific gravity is easily learned by using your hands as scales, a technique known as hefting. Place an unknown mineral in one hand, a known mineral in the other, and while holding your arms slightly away from your body, gently bounce the minerals three or four times in your hands. Then, quickly switch hands and do it again. One mineral should now feel substantially heavier than the other, and the general specific gravity of the unknown mineral can be estimated. Make sure the unknown and known mineral samples are about the same size.

Minerals generally resist breaking, a feature known as **tenacity.** Some are **brittle,** like glass, shattering into many pieces, while others are **malleable,** like many metals, and can be easily deformed. A few are **flexible** (like the easily bent micas), and even more rarely flexible minerals are also **elastic** (they return to their original shapes after the stress is released, like a metal spring).

Striations are shallow, straight, near-microscopic grooves on cleavage surfaces. Orient the specimen so that the flat surfaces reflect light, and striations will become evident. This is a very useful feature to distinguish between the two major groups of feldspars.

Plagioclase feldspars (generally white, cream, gray, or dark in color) show well-developed **striations** on some cleavage surfaces. The striations are very perfectly developed and are precisely spaced or in groups of parallel sets. These grooves will be parallel to the other cleavage face (Figure 1.10).

FIGURE 1.10 Labradorite feldspar with striated surface.

INSIGHTS

Lines resembling striations will appear on the external surfaces of some crystals. These lines are often the result of pauses in the growth of those crystals and do not appear on the cleavage surfaces.

FIGURE 1.11 Nonstriated orthoclase feldspar.

FIGURE 1.12 Striated plagioclase feldspars.

Orthoclase (potassium) feldspars are generally brighter in color than plagioclase types (pink, orange, blue, and green) and lack striations (Figure 1.11 and 1.12).

Effervescence (fizzing) results when a weak solution (3–5 percent) of hydrochloric acid is dropped onto the surface of some carbonates. Calcite, the main mineral present in limestone, reacts well (see Figure 1.13). Dolomite, another carbonate, only fizzes well when the mineral is powdered. When testing for dolomite, the hydrochloric acid will effervesce on ordinary surfaces, but the bubbles will be produced very slowly, while the fizzing on calcite is immediate.

The acid should be supplied in dropper bottles and when used as directed, will not be dangerous. Try not to get it on clothing (it might change the color of dyes) and keep

FIGURE 1.13 Effervescence of hydrochloric acid on calcite.

it off your lab manual (it will degrade the paper). If spilled, just wipe it up. Good manners dictate that you should always wash acid off the minerals you have tested.

Mineral color would seem to be one of the very best tools in identification. Instead, it is one of the most treacherous for the beginner. Some common minerals such as quartz occur in a wide variety of colors. Color is often the result of trace amounts of chemicals within the main mineral. While the different colored varieties of quartz have been used as semiprecious gemstones, they are still only quartz. Calcite and fluorite display a similar wide range of colors (Figures 1.14 and 1.15).

FIGURE 1.14 Color varieties of calcite.

FIGURE 1.15 Color varieties of fluorite.

IN_SIGHTS_

Reserve the use of color until you are nearly finished with the identification process and employ it mainly as a confirming feature, never as a primary character.

Streak color is much more reliable in identifying minerals and is one of your best tools. Streaking is done by dragging the mineral firmly across the surface of a piece of unglazed porcelain (a **streak plate**). Blow off the excess powder and small broken grains and examine the color produced (Figure 1.16).

The color on the streak plate is the actual color of the mineral. The way your eye perceives color is a complex effect. Colors you see are a combination of reflected and absorbed light frequencies, plus the texture of an object will also affect the way you see the color. By rubbing the mineral on a streak plate, the mineral is ground into small particles of uniform size, and therefore all varieties of the same mineral will have identical colors when viewed after streaking.

FIGURE 1.16 Streak plate and color varieties of hematite.

Streak color is a powerful tool. Which samples:

SHOW Colored Streaks	SHOW White Streaks
Sample Number	Sample Number
_________	_________
_________	_________
_________	_________
_________	_________
_________	_________
_________	_________
_________	_________
_________	_________

INSIGHTS

Be sure to press the mineral FIRMLY against the streak plate. If the mineral is only timidly pressed against the porcelain, it may produce a streak so weak that the true color will not be apparent. Remember, STREAK the mineral firmly onto the streak plate, but don't STRIKE the plate.

In some cases, such as with the mineral hematite, the color on the streak plate may be quite different than the apparent color of the unstreaked mineral.

Magnetism occurs naturally in a very few minerals, especially the iron mineral magnetite. This form of the mineral is called a lodestone and will attract iron filings, paper clips, and other magnets. This feature, plus the black streak, is definitive for the mineral. See Figure 1.17.

FIGURE 1.17 Magnetite showing natural magnetism.

FIGURE 1.18 Double refraction of calcite.

Double refraction is a condition where the atomic structure causes light to be broken into two rays that pass through the mineral in different pathways. If such a mineral is transparent, you will see a pair of images when an object is viewed through it. See Figure 1.18.

Some minerals will be **flexible** in thin sheets. Flexible minerals will bend, but after being bent, will not return to their original shape when the bending force is removed. Gypsum is actually flexible in THIN sheets (but the sheets must be very thin and bent very slowly). Other minerals can be both **flexible** and **elastic,** such as micas, which bend and then return to their original flat shapes when the force is removed.

A few minerals will separate easily into fibers such as asbestos. Due to the current controversy over the potential health hazards of this mineral, you will not work with it in your kit, but perhaps your instructor will demonstrate this interesting feature while taking proper safety steps (exhibit fibrous asbestos only in a vented fume hood).

IDENTIFICATION KEY TO MINERALS

METHODS:

(Use the accompanying data sheets to record your observations.)

Identification of minerals is easy when keys are used. The keys presented here (Figures 1.19 and 1.20) use the techniques and physical characteristics described earlier.

STEP ONE: If your unknown mineral shows cleavage, use the first key (Figure 1.19). Start by trying to produce a colored streak, and if you are successful, compare the streak with the colors indicated on the key.

STEP TWO: If there is no streak produced (or if the streak is white), determine the number of cleavages present and the angles between adjacent cleavage planes to find the correct name of the mineral.

STEP THREE: If your unknown mineral does not show cleavage, move to key two (Figure 1.20) and follow the same steps. First, check for streak color and, if unsuccessful, move to the lower part of the key where minerals are differentiated by their hardness.

STEP FOUR: After you feel the mineral has been identified, check Tables 1.1 through 1.5 and see if the detailed clues also match. These tables are arranged by characters: luster, hardness, streak, general properties, and chemical formulae.

Your instructor will provide you with the basic tools for mineral identification:

- Set of 15 to 25 common mineral specimens
- Dropper bottle with 5 percent hydrochloric acid
- Copper penny
- Carpenter's nail
- Masonry nail
- Sheet of glass
- Streak plate

Work in teams of two to four students and use the keys on the following pages to identify the mineral specimens.

NOTE: Be alert to the results of each of your test steps. If you have made an error and are on the wrong chart, you will quickly begin getting nonsense results. If you believe this is happening, turn to the other chart. To a remarkable degree, the chart is self correcting . . . if you pay attention!

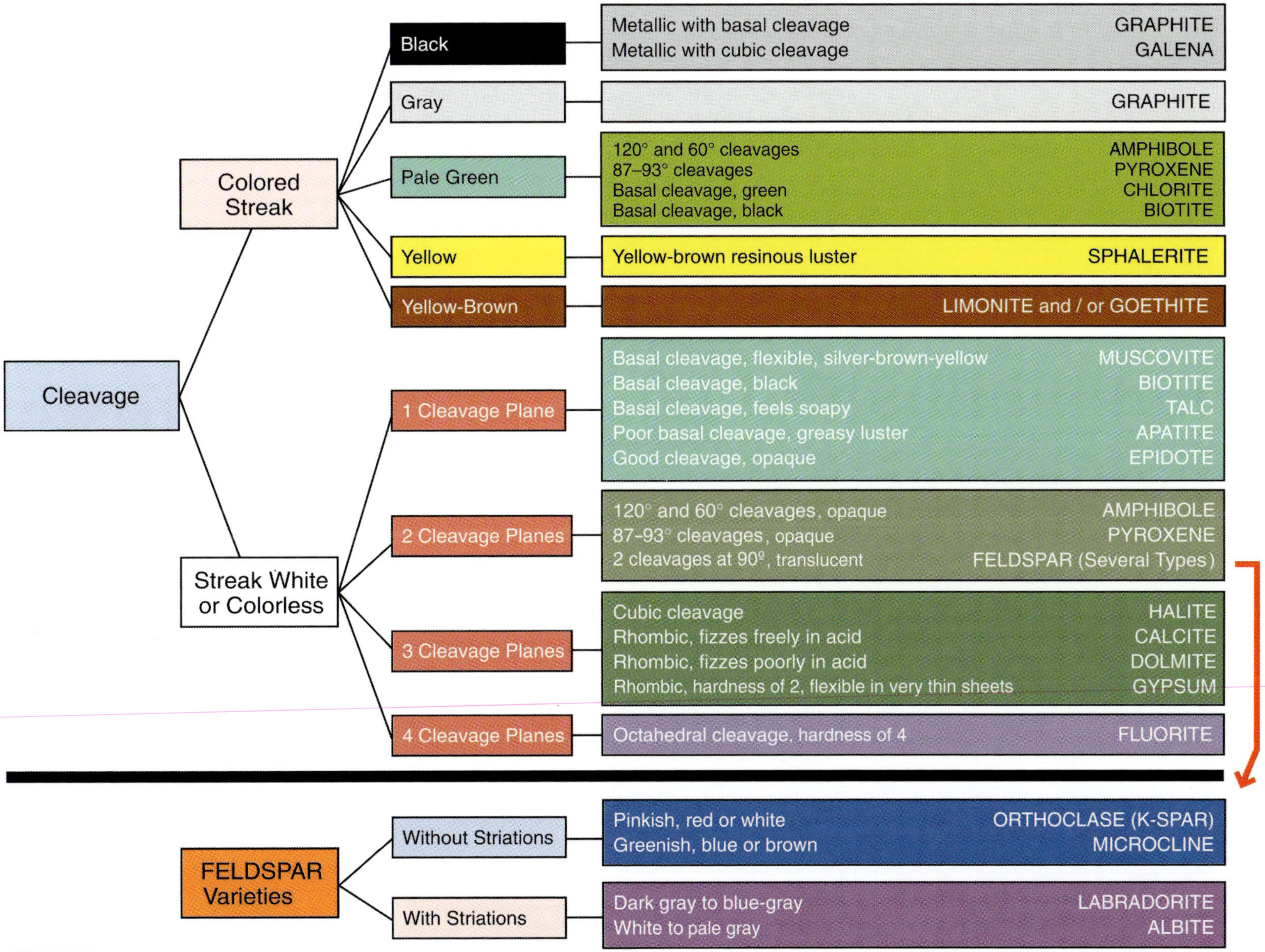

FIGURE 1.19 Minerals WITH Cleavage.

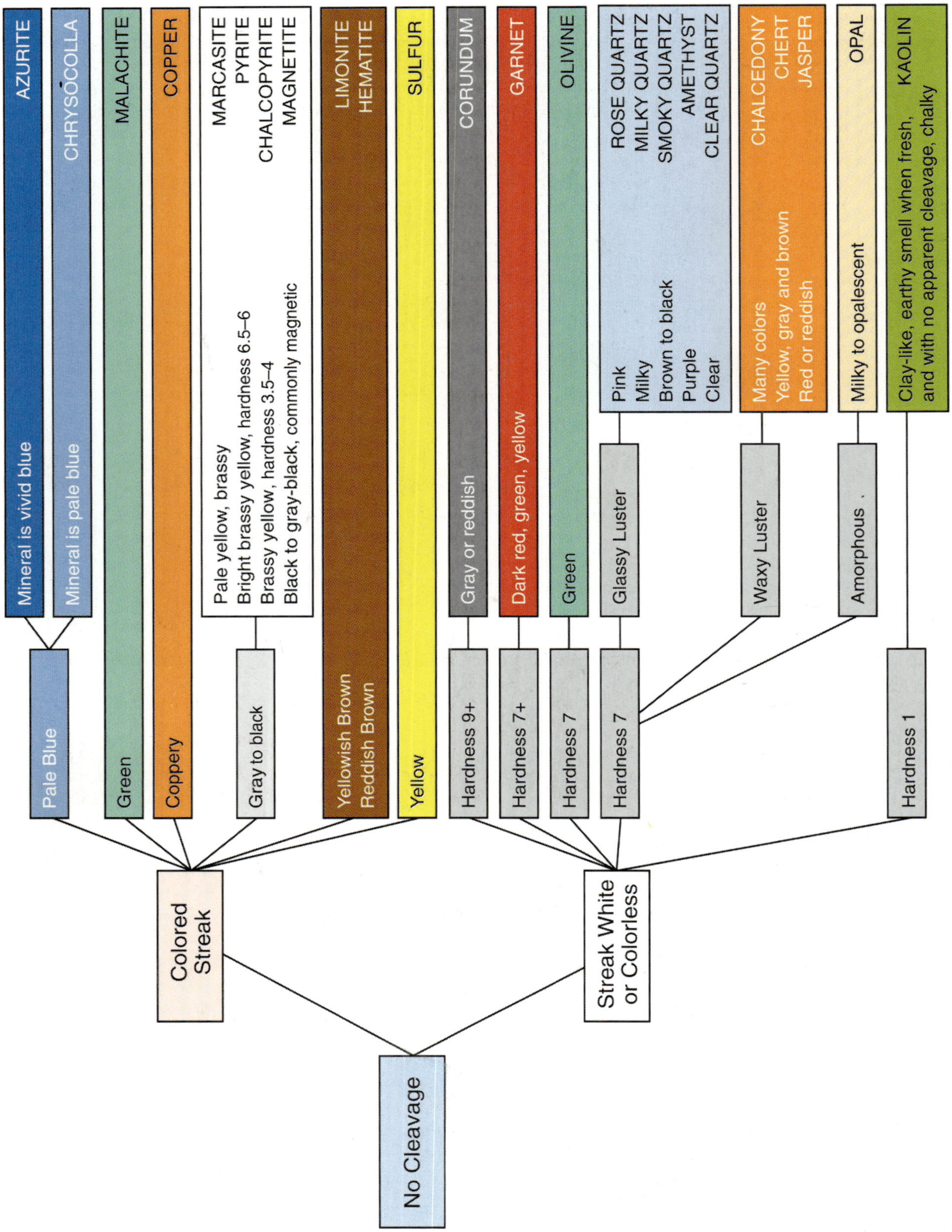

Minerals WITHOUT Cleavage

No Cleavage

Colored Streak

- Pale Blue
  - AZURITE — Mineral is vivid blue
  - CHRYSOCOLLA — Mineral is pale blue
- Green
  - MALACHITE
- Coppery
  - COPPER
- Gray to black
  - MARCASITE — Pale yellow, brassy
  - PYRITE — Bright brassy yellow, hardness 6.5–6
  - CHALCOPYRITE — Brassy yellow, hardness 3.5–4
  - MAGNETITE — Black to gray-black, commonly magnetic
- Yellowish Brown / Reddish Brown
  - LIMONITE
  - HEMATITE
- Yellow
  - SULFUR

Streak White or Colorless

- Hardness 9+ — CORUNDUM — Gray or reddish
- Hardness 7+ — GARNET — Dark red, green, yellow
- Hardness 7 — OLIVINE — Green
- Hardness 7 — Glassy Luster
  - ROSE QUARTZ — Pink
  - MILKY QUARTZ — Milky
  - SMOKY QUARTZ — Brown to black
  - AMETHYST — Purple
  - CLEAR QUARTZ — Clear
- Waxy Luster
  - CHALCEDONY — Many colors
  - CHERT — Yellow, gray and brown
  - JASPER — Red or reddish
- Amorphous
  - OPAL — Milky to opalescent
- Hardness 1
  - KAOLIN — Clay-like, earthy smell when fresh, and with no apparent cleavage, chalky

FIGURE 1.20 Minerals WITHOUT Cleavage.

TABLE 1.1	Metallic Minerals			
	Hardness	Streak	Other Properties	Mineral
Not scratched by steel nail or knife	6.5-6	Dark gray	brass yellow, may tarnish brown; brittle, no cleavage, cubic crystals common, S.G. = 5.0.	PYRITE ("fool's gold") FeS_2 iron sulfide
	6.5-6	Dark gray	pale brass yellow to whitish gold; brittle, no cleavage, radiating masses and "cockscombs," S.G. = 4.9.	MARCASITE FeS_2 iron sulfide
	6	Dark gray	dark gray to black; magnetic, no obvious cleavage, S.G. = 5.2.	MAGNETITE Fe_3O_4 iron oxide
Scratched by steel nail or knife	6.5-5	Red to red-brown	silver to gray, may be tiny glittery flakes may tarnish red, S.G. = 4.9-5.3.	HEMATITE Fe_2O_3 iron oxide
	5.5-5	Yellow-brown	dark brown to black, in radiating layers, S.G. = 4.3.	GOETHITE $FeO(OH)$ hydrous iron oxide
	5.5-5	Yellow-brown	yellow-brown to dark brown; amorphous, but may be pseudomorphic after pyrite, S.G. = 4.1-4.3.	LIMONITE $Fe_2O_3 \cdot nH_2O$ hydrous iron oxide
Scratched by wire nail	4-3.5	Dark gray	golden yellow, may tarnish purple; brittle, no cleavage, S.G. = 4.1-4.3.	CHALCOPYRITE $CuFeS_2$ copper-iron sulfide
	4-3.5	White to yellow-brown	brown to yellow, or black; submetallic, dodecahedral cleavage, S.G. = 3.9-4.0.	SPHALERITE ZnS Zinc sulfide
Scratched by penny	3-2.5	Copper	copper to dark brown, may oxidize green; malleable, S.G. = 8.8-8.9.	NATIVE COPPER Cu copper
	2.5	Gray to dark gray	silvery gray, tarnishes dull gray; cubic cleavage, not scratched by fingernail, S.G. = 7.4-7.6.	GALENA PbS lead sulfide
Scratched by fingernail	1	Dark gray	gray to black, marks paper easily; greasy feel, S.G. = 2.1-2.3.	GRAPHITE C carbon

TABLE 1.2	Nonmetallic Minerals			
	Hardness	Streak	Other Properties	Mineral
Not scratched by steel nail or knife	9	White	gray, red, brown, blue; greasy luster, commonly in six-sided crystals with striated flat ends; no cleavage, S.G. = 3.9-4.1.	CORUNDUM Al_2O_3 aluminum oxide
	8	White	colorless, yellow, blue, or brown; one perfect cleavage, crystal faces often striated, S.G. = 3.5-3.6.	TOPAZ $Al_2O_4(OH,F)_2$ hydrous fluoro-aluminum silicate
	7.5-7	White	green, yellow, pink, blue, brown, or black slender crystals with rounded triangular cross sections; striated crystal faces, no cleavage, S.G. = 3.0-3.2.	TOURMALINE complex silicate
	7	White	any color to colorless, transparent to translucent, greasy luster, no cleavage, conchoidal fracture, S.G. = 2.7.	QUARTZ SiO_2 silicon dioxide
	7	White	white, light colors; waxy luster, translucent, often banded masses cryptocrystalline, S.G. = 2.5-2.8.	CHALCEDONY SiO_2 cryptocrystalline quartz
	7	White	black, cryptocrystalline, waxy, conchoidal fracture, translucent to opaque, S.G. = 2.5-2.8.	FLINT SiO_2 cryptocrystalline quartz
	7	White	gray, brown, yellow; cryptocrystalline, waxy luster, opaque, conchoidal fracture, S.G. = 2.5-2.8.	CHERT SiO_2 cryptocrystalline quartz
	7	White	red, opaque, waxy luster, cryptocrystalline, conchoidal fracture, S.G. = 2.5-2.8.	JASPER SiO_2 cryptocrystalline quartz
	7	White	green, black, or yellow; conchoidal fracture, no cleavage, S.G. = 3.3-3.4.	OLIVINE $(Fe,Mg)_2SiO_4$ ferromagnesian sillicate
	7	White	dark red, brown, pink, green, or yellow; transparent to translucent, no cleavage, S.G. = 3.4-4.3.	GARNET complex silicate

TABLE 1.3			Nonmetallic Minerals	
	Hardness	Streak	Other Properties	Mineral
Not scratched by steel nail, knife	7-6	White	green to yellow-green, striated crystals or dull granular masses, one cleavage, S.G. = 3.3-3.5.	EPIDOTE complex silicate
	6	White	blue-gray, black, or white; striations on some cleavage planes; two cleavages at nearly 90°, S.G. = 2.6-2.8.	PLAGIOCLASE FELDSPAR $NaAlSi_3O_8$ to $CaAl_2Si_2O_8$ calcium-sodium aluminum silicate
	6	White	white, pink, brown, green; exsolution lamellae are present and subparallel, two cleavages at 90°, S.G. = 2.6.	POTASSIUM FELDSPAR $KAlSi_3O_8$ potassium aluminum silicate
	6	White	colorless, white, orange, gray, yellow, green red, blue; may have play of colors (opalescence), amorphous, greasy luster to earthy luster; conchoidal fracture, S.G. = 1.9-2.3.	OPAL $SiO_2 \cdot nH_2O$ hydrated silicon dioxide
Scratched by steel nail, knife	5.5	White	green to black, dull, stout crystals; two cleavage directions that intersect at about 87° and 93°, S.G. = 3.2-3.5.	PYROXENE (AUGITE) calcium ferromagnesian silicate
	5.5	White	green to black, opaque, two cleavage directions at 60° and 120°, slender crystals, may be splintery or fibrous, S.G. = 3.0-3.3.	AMPHIBOLE (HORNBLENDE) calcium ferromagnesian silicate
	5	White	brown, green, blue, yellow, purple or black; one poor cleavage, common as six-sided crystals, S.G. = 3.1-3.2.	APATITE $Sa_5F(PO_4)_3$ calcium fluorophosphate
	5-2	White	green, yellow, gray, or variegated green, gray, and brown; dull masses or asbestos fibrous crystals, no cleavage, S.G. = 2.2-2.6.	SERPENTINE $Mg_6Si_4O_{10}(OH_8)$ hydrous magnesian silicate
	5.5-1.5	Red to red-brown	red, opaque, earthy luster, S.G. = 4.9-5.3.	HEMATITE Fe_2O_3 iron oxide
	5.5-1.5	Yellow-brown	yellow-brown to dark brown, amorphous, but may be pseudomophic after pyrite, S.G. = 3.6-4.0.	LIMONITE $Fe_2O_3 \cdot nH_2O$ hydrous iron oxide

TABLE 1.4	**Nonmetallic Minerals**			
	Hardness	Streak	Other Properties	Mineral
Scratched by wire nail	4	White	colorless purple, blue, yellow, or green; dioctahedral cleavage, crystals usually cubic, S.G. = 3.0-3.3.	FLUORITE CaF_2 calcium flouride
	4-3.5	Light blue	vivid royal blue, earthy masses or tiny crystals, effervesces in dilute HCl, S.G. = 3.7-3.8.	AZURITE $Cu_3(CO_3)_2(OH)_2$ hydrous copper carbonate
	4-3.5	Green	green to gray-green laminated crusts or masses of tiny, granular crystals; effervesces in dilute HCl, S.G. = 3.9-4.0.	MALACHITE $Cu_2CO_3(OH)_2$ hydrous copper carbonate
	4-2.0	Very light blue	pale blue to blue-green crusts or massive, amorphous, conchoidal fracture, S.G. = 2.0-2.4.	CHRYSOCOLLA $CuSiO_3 \cdot 2H_2O$ hydrated copper silicate
	4-3.5	White	white, gray, pink, or brown; opaque; rhombohedral cleavage; effervesces in dilute HCl only if powdered, S.G. = 2.8-2.9.	DOLOMITE $CaMg(CO_3)$ magnesian calcium carbonate
Scratched by penny	3	White	colorless, white, yellow, gray, green, brown, red, blue; transparent to translucent, rhombohedral cleavage; effervesces in dilute HCl, S.G. = 2.7.	CALCITE $CaCO_3$ calcium carbonate
	3	White	colorless, white, red, brown, yellow, blue; platy crystals, massive, or in rose-like shapes; three cleavages, one perfect and at right angles to others; very heavy, S.G. = 4.5.	BARITE $BaSO_4$ barium sulfate
	3	Gray-brown	very dark brown to black, one perfect cleavage; flexible, very thin sheets, S.G. = 2.7-3.1.	BIOTITE MICA ferromagnesian potassium, hydrous aluminum silcate
	2.5	White	colorless, white, yellow, red, blue, brown; cubic crystals and cubic cleavage, salty taste, S.G. = 2.1-2.6.	HALITE $NaCl$ sodium chloride

TABLE 1.5			Nonmetallic Minerals		
	Hardness	Streak	Other Properties	Mineral	
Scratched by fingernail	2.5-1.5	Pale Yellow	yellow to red, bright crystals or earthy masses, brittle, no cleavage, conchoidal fracture, S.G. = 2.1.	NATIVE SULFUR S sulfur	
	2.5-2	White	colorless, yellow, brown, red-brown; one perfect cleavage; flexible, elastic sheets, S.G. = 2.7-3.0.	MUSCOVITE MICA potassium hydrous aluminum silicate	
	2	White	dark green, one perfect cleavage, S.G. = 2.6-3.0.	CHLORITE ferromagnesian aluminum silicate	
	2	White	one good cleavage (two poor cleavages); nonelastic sheets, colorless to white; H = 2, easily scratched with fingernail, S.G. = 2.0-2.4.	GYPSUM $CaSO_4 \cdot 2H_2O$ calcium sulfate	
	2-1	White	white to very light brown, one perfect cleavage, common as earthy, microcrystalline masses, S.G. = 2.6.	KAOLINITE $Al_4(Si_4O_{10})(OH)_8$ hydrous aluminum silicate	
	1	White	white, gray, green, pink, brown, yellow; soapy feel, pearly to greasy luster, massive or foliated, S.G. = 2.7-2.8.	TALC $Mg_3Si_4O_{10}(OH)_2$ calcium carbonate	

■ QUESTIONS FOR FURTHER THOUGHT

1. What minerals do you commonly use daily in their natural form?

2. What common minerals do you recognize in your home?

3. What common minerals were used to make your automobile?

Igneous Rocks

IDENTIFICATION AND PROPERTIES

Igneous rocks are the first of the major rock categories we will investigate. Geologists generally believe that the earth formed in such a way that the first rocks generated by earth processes were all igneous, with sedimentary and metamorphic rocks forming later. As the Rock Cycle demonstrates, all rocks are interrelated, and the concept of **Uniformitarianism** can explain the formation of any type of rock or mineral (Figure 2.1).

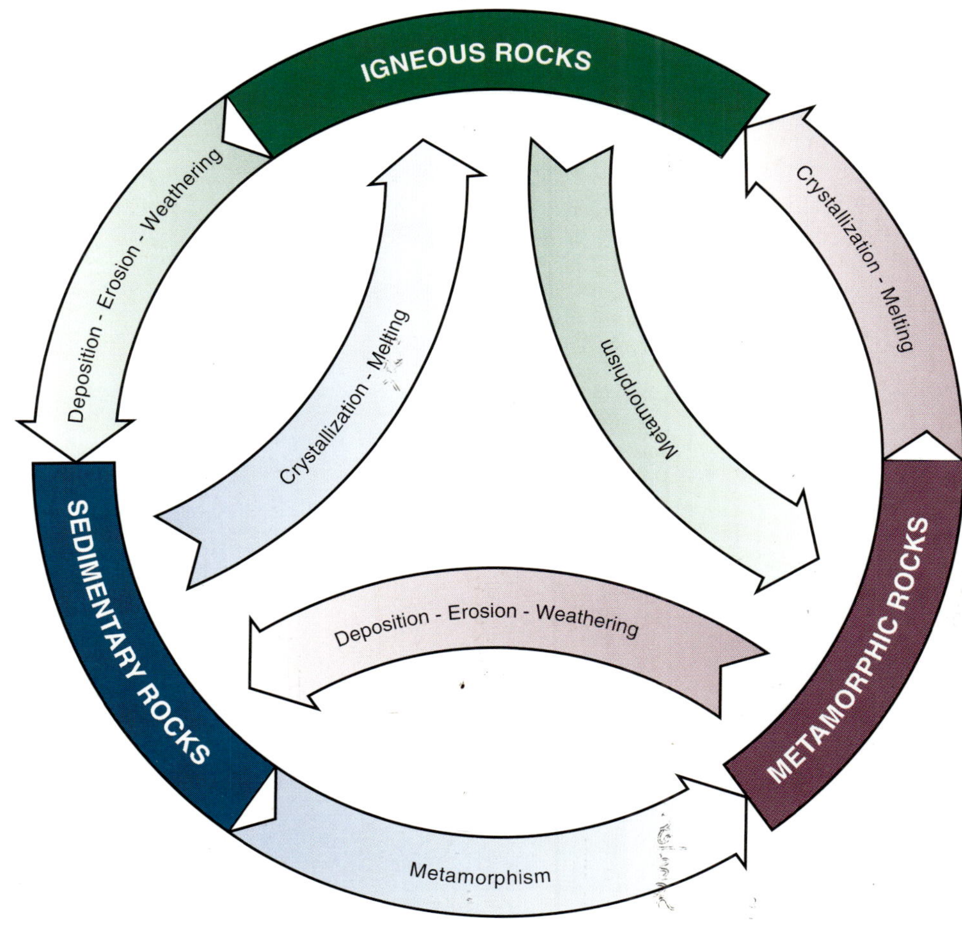

FIGURE 2.1 The rock cycle demonstrates uniformitarianism.

All rocks consist of one or more minerals. Now that you have learned to recognize common minerals, we can move on to identifying rocks.

KEY TERMS

- *Aphanitic Texture*
- *Breccia, Volcanic*
- *Extrusive Igneous Rock*
- *Felsic Igneous Rock*
- *Intermediate Igneous Rock*
- *Intrusive Igneous Rock*
- *Mafic Igneous Rock*
- *Pegmatitic*
- *Phaneritic*
- *Phenocryst*
- *Porphyry*
- *Pumice*
- *Pyroclastic*
- *Scoria*
- *Tuff, Volcanic*
- *Ultramafic Igneous Rock*

Igneous rocks formed during the cooling of melted rock materials **(magma)** originally formed within the earth, melted by pressure and intense heat found there.

Sometimes these molten materials solidified beneath the surface; these rocks are termed *intrusive (coarse grained) igneous rocks.* Once molten magma flows out onto the surface of the earth, it is termed lava, and when crystallized, the resulting rocks will be considered *extrusive (fine grained).* Therefore, crystal size in the final rock is a function of how slowly or how quickly the rock cooled.

■ MATERIALS AND METHODS

- A set of the most common igneous rocks
- A ruler with metric measurements
- A low-powered microscope or a hand lens

Properties of Igneous Rocks

The primary classification of igneous rocks is based on **texture** and color. As stated, intrusive rocks, which cool slowly, are coarse grained, and, if the crystals of the rock are large enough to be seen easily with the unaided eye, the rock is said to be *phaneritic* (Figure 2.2). Sometimes the cooling rate will be unusually slow and the mineral crystals will grow to exceptionally large sizes (crystals on average exceeding one centimeter in size); such rocks are then called *pegmatites* (Figure 2.3).

Extrusive igneous rocks that cool rapidly produce crystals that are too small to be seen and identified by the unaided eye and are termed *aphanitic* (Figure 2.4). If cooled too quickly for crystals to form at all, a volcanic glass is produced, and the texture is said to be **glassy** or **hyaline** (Figure 2.5).

FIGURE 2.2 Phaneritic rock.

FIGURE 2.3 Pegmatitic rock.

FIGURE 2.4 Aphanitic rock.

FIGURE 2.5 Glassy rock.

If the igneous rock is extrusive and gas rich, escaping gases form bubbles that may be trapped in the crystallizing rock. These bubbles are called vesicles, and the resulting rock will appear sponge-like, a condition termed vesicular or frothy (Figure 2.6).

A special case of texture results when a rock records two distinct patterns of cooling and is termed a ***porphyry*** (Figure 2.7). These rocks are primarily intrusive and begin cooling in a manner that produces relatively large crystals ***(phenocrysts)*** and seem destined to become phaneritic rocks. At some point before the rock has entirely crystallized, the temperature drops and the remaining rock mass is formed from aphanitic crystals **(groundmass).** The end product is a rock with both large and small crystals.

FIGURE 2.6 Vesicular rock (scoria).

FIGURE 2.7 Andesite porphyry.

Volcanic tuffs are sometimes difficult to distinguish from rhyolites and andesites that cooled from pools of molten magma (see the Igneous Rock Identification Chart, Figure 2.13). Often, tuffs show layering or rows of tiny bubbles trapped along planes within the rock. Color banding may also be evident in tuffs. Rhyolites and andesites are seldom banded and are generally very uniform in texture.

Two general classes of igneous extrusive rocks are produced during the activity of volcanoes, and the materials produced are termed *pyroclastic.* Often these materials were blasted free from the liquid magma to produce lava that flowed from or was thrown violently out of the crater. If the fragments are very fine (2 mm or less in diameter), they are called *volcanic tuff* or **volcanic ash** (Figure 2.8), but if the grains are larger, they may be encountered as masses of irregularly shaped rock fragments and are then called *volcanic breccias* (Figure 2.9).

Particles of volcanic debris are named according to their general size classes. **Ash particles** are very small (up to 2 mm in diameter, **lapilli** range between 2–64 mm, and pieces of ejected materials larger than 64 mm in diameter are called **blocks** and *bombs* (these can be very large in the case of major volcanic explosion debris).

The other major aid for identification of igneous rocks is general color. Igneous rocks rich in quartz and orthoclase feldspars are light in color (brown, pink, tan, yellow, etc.) and are termed *felsic,* while igneous rocks rich in plagioclase feldspars and **ferromagnesian** minerals are called *mafic* (black, black-brown, greenish-black, etc.). Rocks falling between felsic and mafic commonly have a light and dark speckled appearance and are termed *intermediate.* A fourth class of rocks is the *ultramafics* that consist almost entirely of ferromagnesian minerals.

FIGURE 2.8 Volcanic ash or tuff.

FIGURE 2.9 Volcanic breccia.

WHY ARE THERE SO MANY DIFFERENT IGNEOUS ROCKS?

Bowen's Reaction Series (Figure 2.10) is a very important aid to understanding how different igneous rocks are formed and what governs their chemical content.

Bowen was able to show in the laboratory that, as magmas cooled, two series of silicate rock-forming minerals were produced. One group was the so-called **discontinuous series** where each mineral was only stable under certain temperature and pressure conditions. As the temperature dropped, a series of discrete mineral types formed and disappeared. In the other series, a continuously changing ratio of calcium-rich to sodium-rich plagioclase feldspars formed as the temperature fell **(continuous series).** Near the top of Bowen's chart were the lowest temperature and pressure minerals.

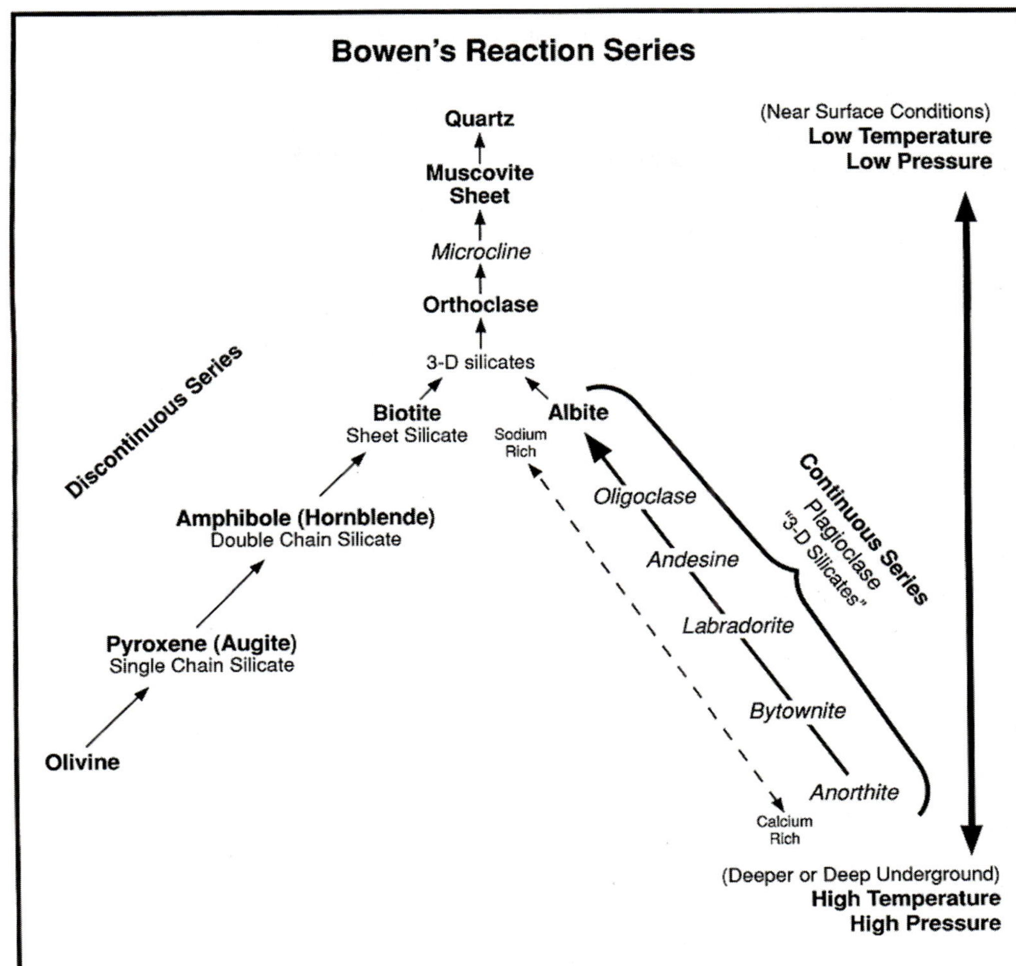

FIGURE 2.10 Bowen's reaction series.

Bowen's Reaction Series demonstrates that, if the magma was cooled while the minerals in the high temperature phase were present, you would get a gabbro (or basalt, gabbro's fine-grained equivalent), while if the series was allowed to run to completion, granite (or rhyolite, granite's fine-grained equivalent) would result. The intermediate case would produce diorite (or andesite, the fine-grained equivalent).

Bowen's chart also demonstrates why you don't find abundant olivine and pyroxene in granite, because these two minerals would not have been present during the formation temperatures that generate low-temperature granitic rocks.

Another possible way to explain the variety of observed igneous rock categories is found in process known as **crystal settling** (or **fractional crystallization**). In this process, a dense mineral formed early in the cooling process might settle out leaving the remaining magma depleted in that mineral. If this occurred, the resulting rock would have a mineral content different from the original magma. This is probably the main reason for the formation of intermediate and felsic magmas (Figure 2.11).

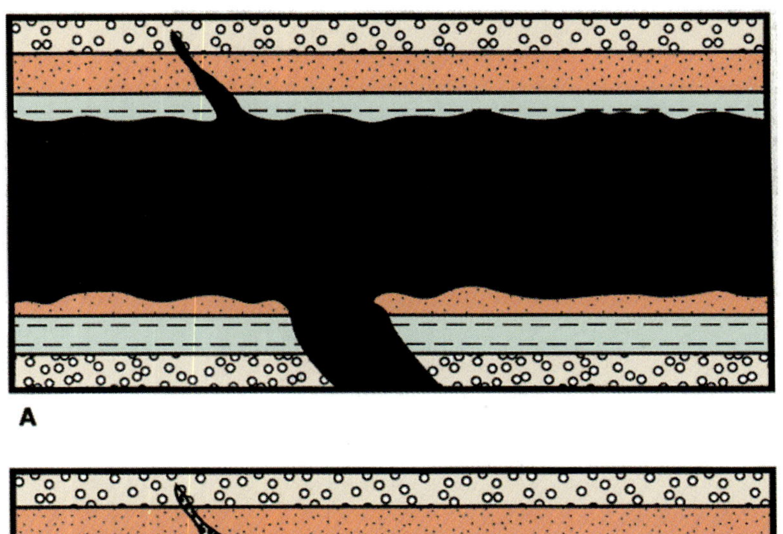

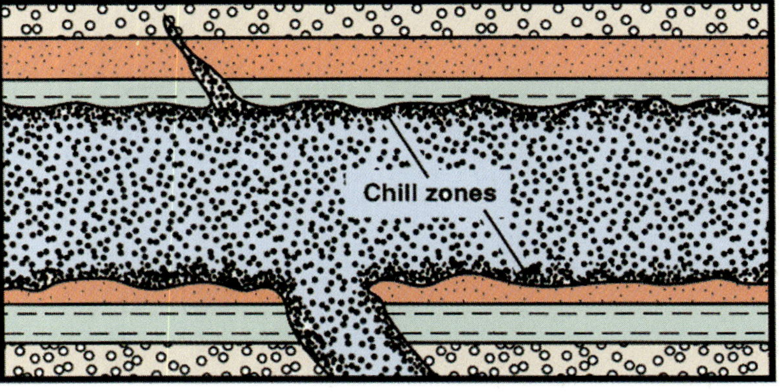

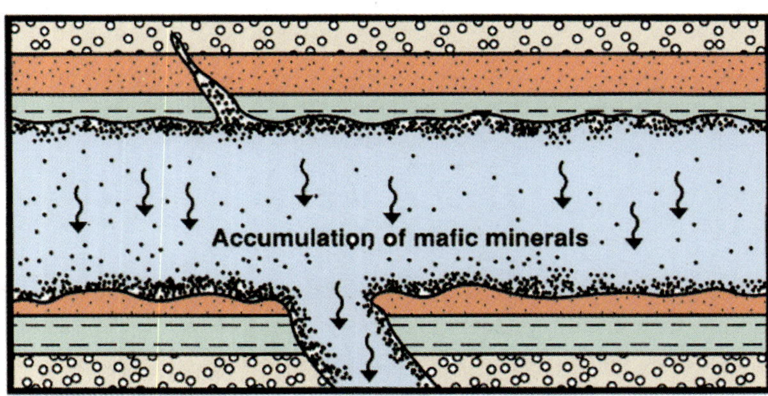

FIGURE 2.11 Crystal settling.

An igneous rock already formed might also be reheated. A glance at the Bowen's chart shows that, if this occurred, the lowest temperature minerals would begin to melt first. In this case, the rock would become partially melted, and the low-temperature minerals could be separated, leaving behind a much more mafic residual rock mass. This process is called **partial melting.**

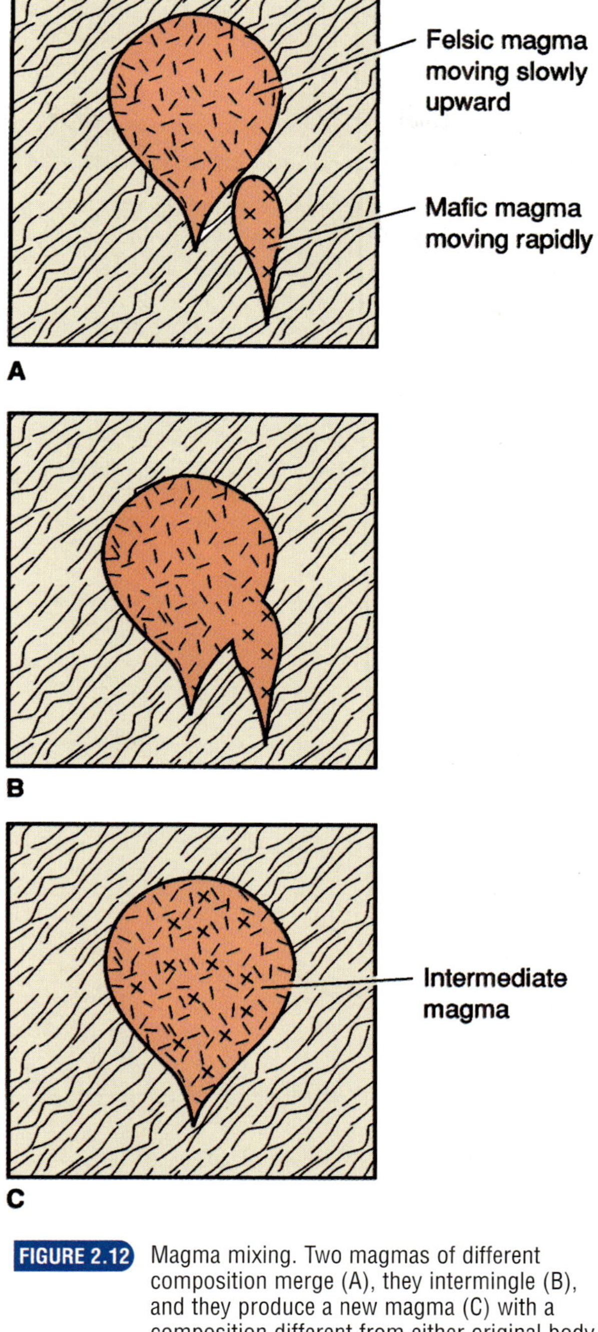

FIGURE 2.12 Magma mixing. Two magmas of different composition merge (A), they intermingle (B), and they produce a new magma (C) with a composition different from either original body.

Although perhaps rare, two magma chambers with different chemical formulations might be joined in the subsurface by means of faults or melting together. If this **magma mixing** occurred, the igneous rock ultimately produced would be different from that in either of the original magmas (Figure 2.12).

Identification Procedures for Igneous Rocks
Using the Identification Key

Black Grain Abundance Chart

Texture \ Composition	Felsic (Light) ≥5% quartz. Potassium feldspar > plagioclase. ≤15% ferromagnesian minerals.	Intermediate <5% quartz. Plagioclase > potassium feldspar. 15–40% feromagnesian minerals.	Mafic (Dark) No quartz. Plagioclase ≤ 50%, no potassium feldspar, ≤40% ferromagnesian minerals.	Ultramafic Nearly 100% ferromagnesian minerals, olivine commonly present and minerals have a high specific gravity.
Pegmatite	GRANITE-PEGMATITE	DIORITE-PEGMATITE	GABBRO-PEGMATITE	PERIDOTITE
Phaneritic	GRANITE (SYENITE)	DIORITE	GABBRO (DIABASE)	
Aphanitic	RHYOLITE (TRACHYTE)	ANDESITE	BASALT	
Porphyry	Granite or Rhyolite Porphyry	Diorite or Andesite Porphyry	Gabbro or Basalt Porphyry	
Glassy	OBSIDIAN			
Frothy or cellular	PUMICE		SCORIA	
Pyroclastic or fragmental	VOLCANIC TUFF (Fragments ≤ 2 mm)			
	VOLCANIC BRECCIA (Fragments ≥ 2mm)			

FIGURE 2.13 Key for the identification of igneous rocks.

 **IN**SIGHTS

For porphyritic textures, simply combine the basic rock name with the terms porphyry or porphyritic. For example, granite with phenocrysts can be called porphyritic granite, or granite porphyry.

Syenite is a name applied to felsic rocks resembling granite, but containing no quartz.

Trachyte is a name applied to felsic rocks that resemble rhyolite, but containing no quartz.

Diabase is a name applied to gabbros composed almost totally of plagioclase and pyroxene mineral crystals about 1.2 mm in size.

Peridotite consisting primarily of olivine is called a **dunite.**

All pale to white mineral grains are NOT quartz. Check with the microscope to make certain that light-colored grains do not show cleavage and striations, making them plagioclase feldspars.

If your rock is PHANERITIC, you can make use of the '**Black Grain Abundance Chart**' above the Identification Chart.

In this strip chart, black dots represent the relative abundance of BLACK grains (crystals) in the rock. Slide your rock sample along the strip chart and look for the relative abundance of truly BLACK grains, not grains that are gray, brown, green or any other color. The relative abundance of the black grains will often lead you directly to the correct color column on the chart.

NOTE: This will not work on aphanitic rocks, because you cannot clearly see how many black grains are present.

IDENTIFICATION KEY TO IGNEOUS ROCKS

METHODS:

(Use the accompanying data sheets to record your observations.)

STEP ONE: Texture of igneous rocks is a primary classification feature. Determine whether your sample is **phaneritic** (coarse grained), **aphanitic** (fine grained), glassy or exhibits some unusual texture.

> **Phaneritic Rocks:** Decide whether your sample is just phaneritic or is coarse enough to be a **pegmatite,** or whether there are two size classes of mineral crystals present making it a **porphyry.**

> **Aphanitic Rocks:** Decide whether your rock is simply fine grained, glassy, **vesicular,** or **pyroclastic.**

STEP TWO: Determine whether the sample is **felsic** (light in color), **intermediate,** or **mafic** (dark in color). **Ultramafic** rocks are not only dark colored but are generally much more dense than expected for their size.

> **Phaneritic Rocks:** Check the minerals present to see if they support your color decision. For example, a felsic rock should have abundant quartz.

> **Aphanitic Rocks:** You can't normally see which minerals are present here and must rely on general color alone.

> **Glassy:** Color is of little use with these rocks. Rely on texture alone.

> **Frothy or Vesicular:** Gray to white colored samples are normally considered **pumice,** while red to black samples are termed **scoria.**

STEP THREE: Determine the relative abundance of **ferromagnesian** grains to all other mineral grains. Ferromagnesian minerals are dark-colored minerals. The strip chart accompanying your key shows black dots (ferromagnesians) in contrast to **ALL** grains of any other color. If you hold your rock sample against the chart and make a visual percentage estimate, it will be one more way to assure yourself that you are in the correct part of the chart.

> **Phaneritic Rocks:**

> > **Felsic** types will have few to no ferromagnesian grains (0–15 percent).

> > **Intermediate** rocks will commonly have a light-to-dark "salt-and-pepper" appearance due to a mix of ferromagnesian and non-ferromagnesian grains (15–40 percent ferromagnesians).

> > **Mafic** rocks will have large percentages of ferromagnesian grains (40 percent or more) giving the rocks their characteristic dark colors.

> > **Ultramafic** rocks will consist almost entirely of ferromagnesian minerals.

> **Aphanitic Rocks:** Since grain size is so small, you will not be able to determine which minerals are present without a microscope.

Results: If you carefully follow these steps, you realize that once you correctly check color and texture, the intersection of the color column and the texture class will provide the name of the rock being examined.

Using the identification key (Figure 2.13), record the data for each of your specimens on the included data pages. If you are careful and follow these instructions, you will find it easy and enjoyable to identify your igneous rock samples.

■ OTHER ACTIVITIES

Internet Links

Check the Internet for links to sites discussing igneous rocks. Use these Internet sources to find information on useful igneous rocks in your locality. Several Internet sites have numerous high-quality illustrations of both common and rare types.

Visit Local Museums

There are many more types of igneous rocks than we have examined here, but the majority of the named commercial building and monumental stones are variations on the ones you have studied.

Urban Field Trips

Your instructor can also be helpful in setting up an urban field trip. Most cities have numerous public and private buildings that use stone in their construction. In many cases, these stones are igneous or metamorphic and are composed of minerals you have studied. Visit your city center or public buildings and see how many minerals you can identify from this chapter's study.

Just for fun, go to a building supply center and see how many different types of stones are being sold as "granite" or "marble," which are actually badly misnamed. See how many you can properly identify, now that you have studied the igneous rocks.

Other Uses

What other uses are there for igneous rocks? Just a few to get you started include: crushed granite is fed to chickens who use the sharp rock fragments in their gizzards to "chew" their food; scoria and pumice are popular landscaping materials; and perlite and vermiculite are important materials to give good growing properties to potting soils.

Ask people in different industries whether they can suggest other igneous rock types used to make your life richer.

Data Sheets for Igneous Rocks

Sample #	Texture	Rock Name	Minerals Seen in Abundance	Color	Other Properties — How did you determine this?

Sedimentary Rock Identification

Sedimentary rocks, one of the three major rock categories (Figure 3.1), are formed in one of several ways:

1. **Compaction** of loose sedimentary particles with or without cementation

2. **Direct precipitation** of crystalline materials during the evaporation of water

3. The **accumulation of organic debris** (shells, wood, etc.)

4. Cementation by one or more chemical cements

In the first case, loose particles of sand or mud can undergo simple compaction until the grains are locked together forming a rock.

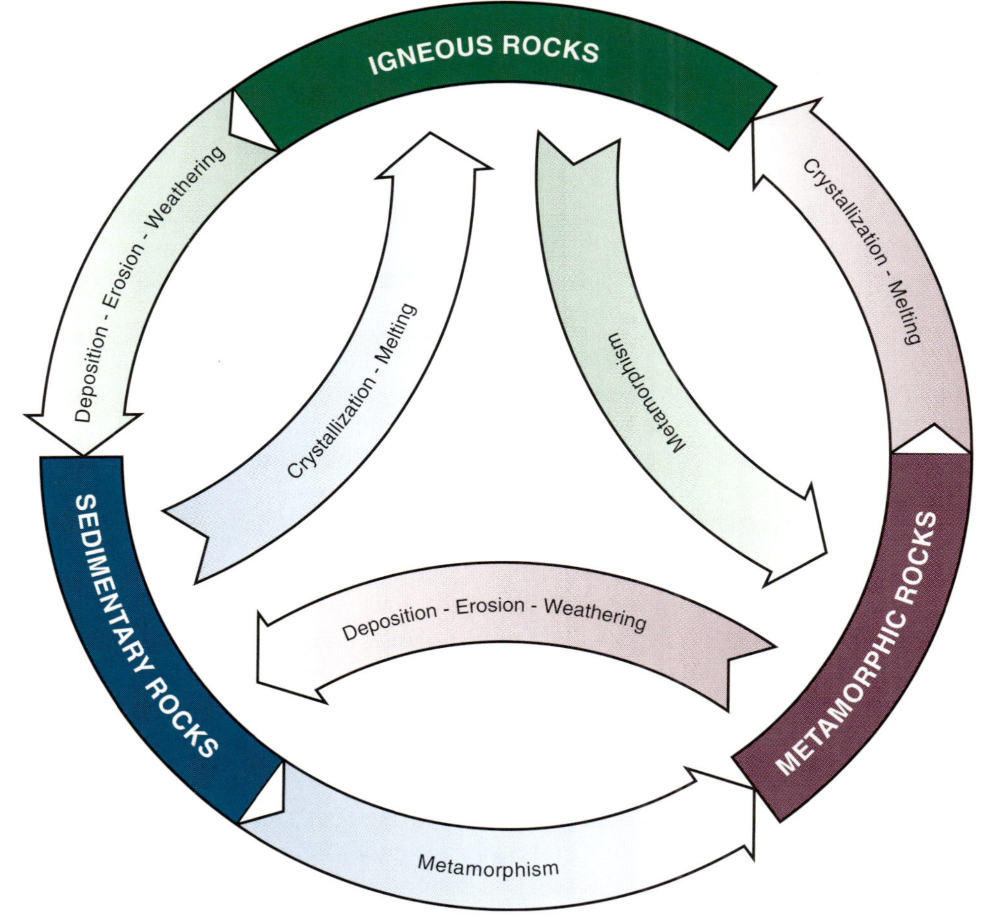

FIGURE 3.1 Geologic rock cycle.

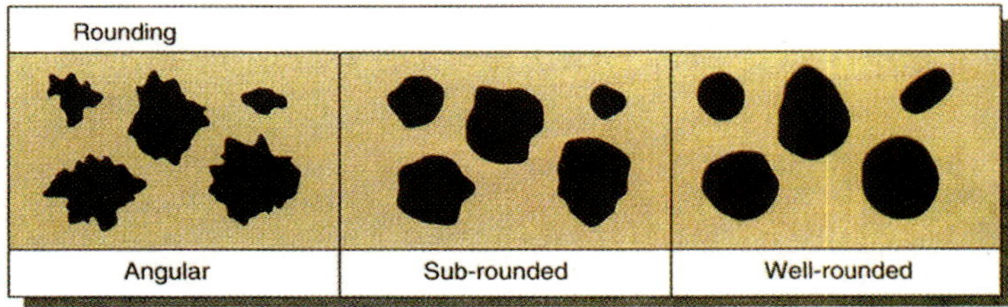

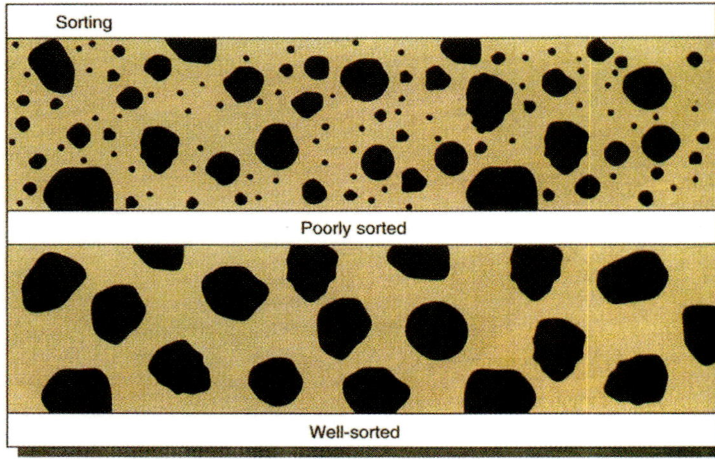

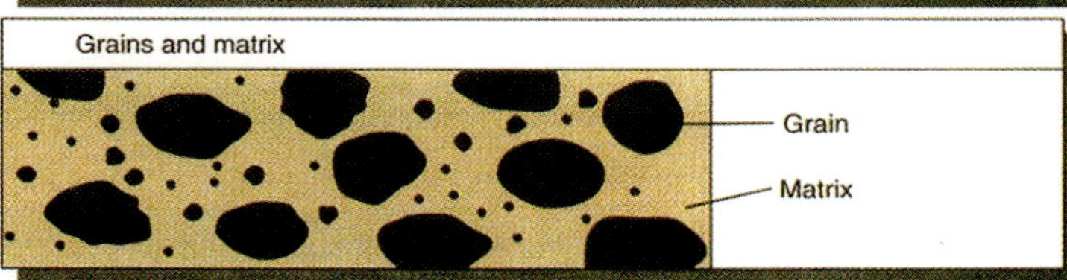

FIGURE 3.2 Rounding and sorting of sediment grains. From *Laboratory Manual for Physical Geology* by Norris Jones. Reprinted by permission of the McGraw-Hill Companies.

The same rock material could be cemented together by the growth of calcite or quartz crystals bridging between the grains and filling the pore spaces. The second case generates minerals like gypsum, halite, and dolostone and chemical limestones where the evaporation process produces a supersaturated condition, forcing the less soluble chemical compounds to form crystals and settle out of the water.

Sediment itself can consist of many different things. Simply stated, sediment often consists of loose and fragmented materials that may either have formed in place or may have been transported from some remote point of origin, its **provenance.** The sedimentary particles were originally produced either by **chemical weathering, mechanical weathering, biologic activities** producing or altering some parent material, or some combination of these processes.

Much sediment is transported, and the physical process involved produces effects known as **rounding** and **sorting** (Figure 3.2). Rounding is the mechanical result of grain-to-grain impact, mechanical weathering, or chemical weathering. Because sharp edges and corners are more easily removed than smooth surfaces, grains become more rounded with further weathering or distance traveled from the source area. This is the main reason that much sand and river rock is round or oval in shape.

In rounding, angular grains retain sharp edges and represent materials close to the primary source material, while subangular and subrounded grains have been altered by weathering and transport to an ever more rounded state. Rounded grains may approach spherical shapes and represent abraded mature sediment transported far from the source terrane.

Sorting is the result of different amounts of depositional energy being available within subdivisions of a transport system. High depositional energy (fast currents and turbulent water) can move large particles while lower energy levels (quiet or slow-moving water) only move the smallest grains. Thus, a river will concentrate coarse sand and gravels in one area while depositing mud elsewhere. Poorly sorted sediments, those containing a mixture of grain sizes and shapes, are generally found close to the source area while well-sorted sediments may consist of a single grain type and shape due to winnowing and transport (Figure 3.2).

Another important feature of sediments is the grain size. Geologists use the **Wentworth Scale** to describe the sizes present in a sample (Figure 3.3).

Size range (metric)	Size range (approx. inches)	Wentworth Class
> 256 mm	>10.1 in	Boulder
64–256 mm	2.5–10.1 in	Cobble
32–64 mm	1.26–2.5 in	Pebble, very coarse gravel
16–32 mm	0.63–1.26 in	Pebble, coarse gravel
8–16 mm	0.31–0.63 in	Pebble, medium gravel
4–8 mm	0.157–0.31 in	Pebble, fine gravel
2–4 mm	0.079–0.157 in	Granule, very fine gravel
1–2 mm	0.039–0.079 in	Sand, very coarse
1/2–1 mm	0.020–0.039 in	Sand, coarse sand
1/4–1/2 mm	0.010–0.020 in	Sand, medium sand
125–250 microns	0.0049–0.010 in	Sand, Fine sand
62.5–125 microns	0.0025–0.0049 in	Sand, very find sand
3.90625–62.5 microns	0.00015–0.025 in	Silt Mud
< 3.90625 microns	< 0.00015 in	Clay Mud

FIGURE 3.3 Non-graphical Wentworth Scale

MAJOR CLASSES OF SEDIMENTARY ROCKS

Clastic Sediments and Detrital Rocks

Clastic materials are those primarily broken away from some parent material by the process of mechanical weathering, and a *clast* is one such fragment (sand, rocks, plant debris, shells or bones).

We have already discussed **pyroclastic** rocks (see Chapter 2), but there are two additional types geologists commonly recognize: **siliciclastic** and **bioclastic.** Siliciclastic sediments consist of the weathering debris from silicate rocks—that is, clay, silt, sand, gravel, or fragments of silicate rocks. Bioclastic sediments are those composed of fragments of once-living organisms. Bioclastic sediments can consist of such diverse materials as fragments of wood and leaves, animal bones, and shells.

Detritus is a collective term for loose rock and mineral material worn off and removed directly by mechanical means such as sand, silt, and clay derived from older rocks and removed from its place of origin. It is normally considered to consist of clay minerals, fresh and abraded mineral grains, and fragments of preexisting rocks. Rocks formed from these sediments are considered detrital rocks (or clastic rocks).

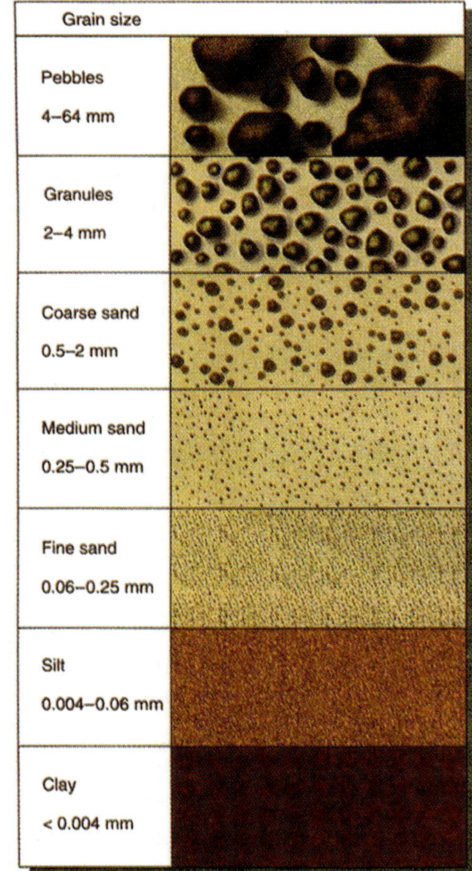

FIGURE 3.4 Wentworth Grain Size Scale. From *Laboratory Manual for Physical Geology* by Norris Jones. Reprinted by permission of the McGraw-Hill Companies.

Some geologists have named such fine-grained rocks as **claystones**, but this usage is only proper when the materials are known to be clays—that is, a clay is one of a suite of closely related sheet-silicate minerals. This fact is seldom known and the fine-grained rock may well be composed of very fine particles of other siliciclastic debris. Current usage suggests that fine-grained rocks be called **"mudstones"** when the chemical nature of the sediment is unknown, and that the use of the term "claystone" should be restricted to rocks where clay minerals are known to be the principal siliciclastic material present (clay making up 80 to 90 percent of the rock mass).

Clastic rocks are further classified mainly on their grain size and shape. Common classes are taken largely from the **Wentworth Scale** and are: clay, silt, sand, gravel, and larger particles.

Rock names are determined by the size of their particles and to some degree by the process of their formation. Hence, a rock composed of very fine clay-sized particles would be a *mudstone* if it is poorly compacted and breaks into rough blocky chunks, but the same material would be a *shale* if it broke into flat sheets and could then be said to be *fissile* (Figure 3.5).

Other siliciclastic rocks are named according to their proximity to the source area and according to the presence of key minerals. Examples of these are the rock type *arkose,* which is dominated by unweathered and weathered orthoclase feldspar grains; **lithic sandstone,** which contains grains that are recognizable rock fragments; and *graywacke* (also referred to as **wackes** and **wackestones**), which are quartz sandstones containing abundant, generally dark-colored muds and clays (hence a "dirty sandstone").

Shape of the rock particles is also important. **Breccia** is defined here much as it was in the last chapter (see volcanic breccia in Chapter 2). Rocks composed of irregular, angular, unsorted debris can form a *sedimentary breccia,* while *conglomerates,* which may have similar sized fragments, are composed of rounded grains (Figure 3.6).

FIGURE 3.5 Mudstone and shale.

Nearly all clastic sedimentary rocks will scratch glass, since they generally contain quartz, but with the exception of chert, almost no chemical sedimentary rocks are as hard as glass.

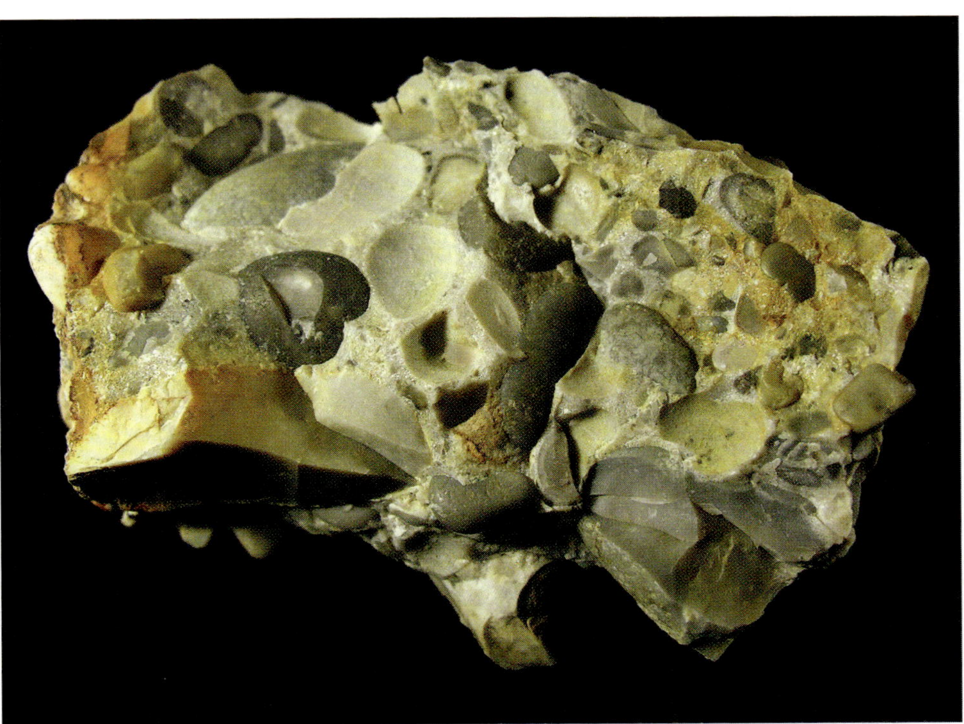

FIGURE 3.6 Sedimentary breccia (top) and conglomerate (bottom).

FIGURE 3.7 Coquina (left) and fossil-rich calcirudite (right).

Chemical and Biochemical Sedimentary Rocks

Chemical sedimentary rocks are those that are entirely produced by chemical means. These are crystalline materials produced directly from water (calcite, aragonite, dolomite, halite, gypsum, chert, opal, etc.). The resulting rocks are termed chemical limestone when the minerals are calcite or aragonite, dolostone when the carbonate is dolomite, rock salt when the material is halite, rock gypsum when it is gypsum, and chert when the material is microcrystalline quartz.

The rock known as **_dolostone_** is formed from the mineral dolomite, either directly precipitated from seawater or by later modification of chemical limestones by the addition of magnesium, altering the formula from $CaCO_3$ (calcite) to $CaMg(CO_3)_2$ (dolomite).

The sedimentary limestone properly termed **travertine** is an important chemical sedimentary rock often seen in building interiors as floors and wall facings, in stone "eggs" imported from Mexico, and in carved "alabaster" bookends. In the trade, this form of travertine is often called "Mexican onyx" or "alabaster." True alabaster is a rock formed entirely from the mineral gypsum, referred to in this manual as **rock gypsum.**

Biochemical sedimentary rocks consist largely or entirely of material that was once living. Examples of these materials include coal, **_coquina_** (comprised nearly entirely of whole and broken shells), and **_calcirudites_** (rocks containing coarse sand to gravel-sized particles, with large percentages of broken or whole shells) (Figure 3.7).

Miscellaneous Sedimentary Rocks

A few common rocks do not fall under the clastic or carbonate classification schemes, and these are listed in Figure 3.10. If your sample does not effervesce (fizz) and will not scratch glass, it probably belongs in Figure 3.10. The only exception is **_chert,_** which is microcrystalline quartz and will easily scratch glass. Chert may be found in many colors, but it always has the look and feel of porcelain.

By now you should be able to easily recognize halite and gypsum, while peat and coal will be dark and obviously organic in nature. **Peat** is the partially decomposed remains of dead plants, and sedimentary coal, termed **bituminous coal,** is the result of increased pressure and age on peaty materials. Peat is generally brown, but bituminous coals are shiny and black.

IDENTIFICATION KEY TO SEDIMENTARY ROCKS

METHODS:

Identification Procedures and Techniques

The keys presented here are arranged in three parts: (1) siliciclastic sedimentary rocks; (2) carbonate sedimentary rocks; and (3) miscellaneous sedimentary rocks. The steps below should allow you to quickly identify your unknown rock samples:

1. Attempt to scratch a glass plate with the sample. If it scratches the glass, it WILL either be a **siliciclastic** rock or chert. Use Figure 3.8 if the sample is granular; use Figure 3.10 if the sample scratches glass easily but has a smooth, nongranular texture.

   a. These samples will scratch the glass only if the grains are sand sized or larger. Examine the grains, determine their size and shape, and move to the siliciclastic chart.

   b. Some **siliciclastic** rocks are too fine grained to produce easily observed scratches (*siltstones* and mudstones). Check these in step 4 below.

2. If the rock will not scratch glass, put a drop of dilute hydrochloric acid on the sample. If it **effervesces** (fizzes), it will be EITHER a **carbonate rock** or a **carbonate cemented siliciclastic** rock (if the latter, it should have scratched glass). Use Figure 3.9.

   a. Determine if the rock is well-cemented and dense or poorly cemented and porous.

      —If well cemented and dense, with visible fossil fragments present, determine the grain size and consider **calcirudite,** *calcarenite, calcisiltite,* or *micrite.*

      —If well cemented and dense, coarsely crystalline without evident fossil fragments being present, determine the grain size and consider **crystalline limestone,** or **micrite.**

      —If poorly cemented and earthy, with visible fossil fragments present, consider **coquina** or **chalk,** depending on the grain size present.

   b. If the rock is composed of small spherical laminated grains, consider an *oolitic limestone* (small, spherical, concentrically laminated grains).

   c. **Dolostone** can be tricky. Either scratch the surface before applying the acid or pulverize a small bit of the rock. If the acid is applied to the surface of your sample, it may take as long as 5–10 seconds for the effervescence to begin.

3. Figure 3.10 covers those miscellaneous, common chemical sedimentary rocks. These samples will not effervesce and, with the exception of chert, will not scratch glass. Use the same tests here that you used for those minerals in Chapter 2.

CAUTION

Don't be in too great a hurry with the acid testing. If the carbonate is **dolostone,** the effervescence will not start immediately, and it will not be as evident as that seen for calcite rocks (see section C).

IN SIGHTS

Put the acid in a crack or hole in the rock to increase the surface area in contact with the acid.

INSIGHTS

Put a drop of acid on the unknown rock sample and wait until all fizzing stops. Take a clean paper towel and firmly blot the spent acid (don't wipe or scrub the rock). Examine the wet area of the paper towel. If it is stained by mud the same color as the rock sample, then the rock was cemented and the rock grains were free to move to the paper towel as a pigment.

If the wet area on the paper towel is more or less colorless, the rock was a chemical limestone, and BOTH the cement and the rock grains were dissolved by the acid.

4. Very fine-grained sedimentary rocks can be a challenge. If grain sizes are silt-sized or smaller, test with acid. If they effervesce, they are **PROBABLY** micrites, chalks, or calcisiltites. If they **DO NOT effervesce,** they **MUST** be siliciclastics and will therefore be mudstones, siltstones, or shales.

Problems arise because some siliciclastics are cemented by calcite and will effervesce when the acid acts on the cement, but not on the rock components.

Once you have decided that the sample is a fine-grained siliciclastic rock, you can easily determine whether its grain size is silt- or clay-sized. With the assistance of your instructor, remove a sample NO LARGER than a pinhead and crush it gently between your incisors [front teeth]. If it crushes to a smooth paste with a few nibbles, the rock was a mudstone or muddy (argillaceous) shale. If it remains slightly gritty, it was a siltstone or silty shale.

Silly as this test may sound, it is the only really useful field technique for a geologist trying to determine the particle size of fine-grained rocks. Don't be afraid of "eating rocks." Give it a try! Just nibble gently and use a SMALL fragment!

■ OTHER ACTIVITIES

Internet Links

Check the Internet for links to sites discussing sedimentary rocks. Use these Internet sources to find information on useful sedimentary rocks in your locality. Several Internet sites have numerous high-quality illustrations of both common and rare types, plus much information on the formation of these interesting rocks.

Urban Field Trips

Your instructor can also be helpful in setting up an urban field trip like that discussed in the last chapter. Most cities have numerous public and private buildings that use stone in their construction. In many cases, these stones are igneous or metamorphic and are composed of minerals you have studied, but many are also constructed of sedimentary rocks. Visit your city center or public buildings and see how many types you can identify from this chapter's study.

Composition	Grain-size Class and Diameter		Comments	Name	
Mainly quartz, feldspar, rock fragments, and clay minerals	gravel (>2mm)		rounded grains	CONGLOMERATE	
			angular grains	BRECCIA	
	sand (0.0625-2.00, or 1/16-2, mm)		mostly quartz grains	QUARTZ SANDSTONE	SANDSTONE
			mostly feldspar grains	ARKOSE	
			mostly rock fragments	LITHIC SANDSTONE	
			mixed with much silt and clay	GRAYWACKE	
	mud and silt	silt (0.0039-0.0625, or 1/256-1/16, mm)	nonfissile (compact)	SILTSTONE	MUDSTONE
			fissile (splits easily)	SILTY SHALE	
		mud (<0.0039, or 1/256, mm)	nonfissile (compact)	MUDSTONE	
			fissile (splits easily)	MUDDY SHALE	

FIGURE 3.8 Chart I: siliciclastic rocks.

Composition	Comments	Grain-size	Name	
Mainly calcium carbonate, CaCO$_3$	shells or shell fragments (i.e., skeletal grains) well cemented to form dense rock	gravel (> 2 mm)	CALCIRUDITE	SKELETAL LIMESTONE
		sand (0.0625-2 mm)	CALCARENITE	
		silt (0.0039-0.0625 mm)	CALCISILTITE	
		clay (< 0.0039 mm)	MICRITE	
	shells or shell fragments (i.e., skeletal grains) poorly cemented to form porous, earthy rock	gravel (> 2 mm)	COQUINA	
		sand (0.0625-2 mm)	CALCARENITE	
		silt and clay (<0.0625 mm)	CHALK	
	spherical grains with concentric laminations	< 2 mm	OOLITIC LIMESTONE	
	crystals formed as inorganic chemical precipitates	coarse-grained (> 2 mm) to fine-grained (0.0039 mm)	CRYSTALLINE LIMESTONE	CHEMICAL LIMESTONE
		very fine-grained (< 0.0039 mm)	MICRITE	
Mainly dolomite, CaMg(CO$_3$)$_2$	commonly altered from limestone	all sizes	DOLOSTONE	

FIGURE 3.9 Chart II: carbonate rocks.

Composition	Comments	Grain-size	Name
Mainly varieties of quartz, SiO_2 (chalcedony, flint, chert, opal, jasper, etc.)	Commonly occurs as layers, lenses, nodules	microcrystalline or amorphous	CHERT
Mainly halite, NaCl	crystals formed as inorganic chemical precipitates	all sizes	ROCK SALT
Mainly gypsum, $CaSO_4 \cdot 2H_2O$	crystals formed as inorganic chemical precipitates	all sizes	ROCK GYPSUM
Mainly plant fragments	brown and porous	all sizes	PEAT
	black and nonporous	all sizes or dense with conchoidal fracture	BITUMINOUS COAL

FIGURE 3.10 Chart III: miscellaneous sedimentary rocks.

Data Sheets for Sedimentary Rocks

Sample #	Rock Name	Texture	Composition	Other Properties How did you determine this?

Data Sheets for Sedimentary Rocks

Sample #	Rock Name	Texture	Composition	Other Properties How did you determine this?

Data Sheets for Sedimentary Rocks

Sample #	Rock Name	Texture	Composition	Other Properties / How did you determine this?

Data Sheets for Sedimentary Rocks

Sample #	Rock Name	Texture	Composition	Other Properties How did you determine this?

4

Metamorphic Rock Identification

The third major group of rocks are the *metamorphic* types. The term metamorphic is derived from a pair of Greek words and means altered or changed **(meta)** in form **(morph).** The change or alteration is generally expressed by the appearance of a new set of minerals, new crystal shapes, new textures, and often by the generation of rocks that are tougher and more stable than the original rocks.

Metamorphic processes alter the preexisting rocks, which are known as the **precursor** materials. These precursor rocks may have begun their existence as either igneous, sedimentary, or even other metamorphic rocks (review the **Rock Cycle,** Figure 4.1). During the metamorphic process, these original materials may be so severely altered that they become unrecognizable.

The agents of metamorphism that cause these sometimes small, sometimes profound changes are simple: heat, pressure, time, and the presence of fluids (often simply water).

The metamorphic process proceeds in part through heat generated by one of several geologic processes. In one case, the heat produced as the two sides of a fault slide past each other will produce metamorphic change, although the effects are limited to nearby rocks **(fault zone metamorphism).**

Contact metamorphism (Figure 4.2) is a much more powerful agent of change through the contact of intruding igneous bodies that bake the surrounding rocks, flood them with hot chemical fluids, and apply considerable pressure. These fluids and chemicals escaping from the contact zone often produce complex mineral assemblages by a process called **hydrothermal alteration,** frequently depositing important metallic ore minerals.

The third major type of alteration is termed **regional metamorphism,** a much larger process involving changes over large regional areas, sometimes entire mountain ranges. The forces in this largest of metamorphic processes come from the pressure and heat caused by folding, faulting, and deep burial of rocks during mountain-building events **(orogenies).**

Further changes may come from very large igneous intrusions producing effects like those of contact metamorphism, but over a large area. Finally, regional metamorphism may cause large-scale migrations of chemicals and fluids that further alter the regional rock masses. Plate tectonics and the resulting collisions are often the provider of energy to drive these three processes.

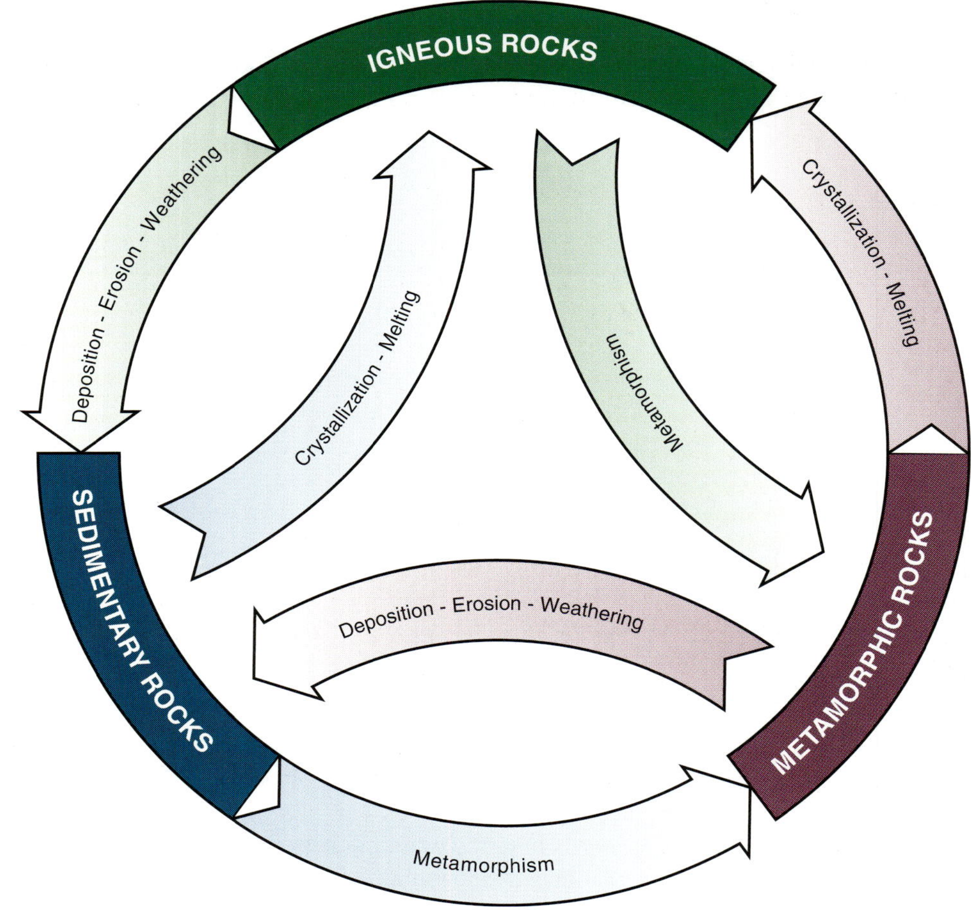

FIGURE 4.1 Rock cycle.

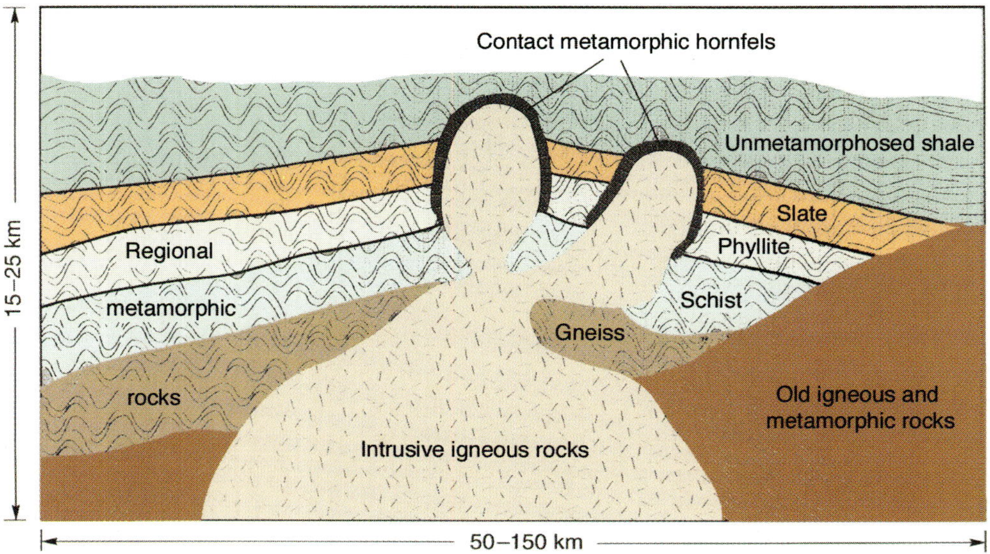

FIGURE 4.2 Contact metamorphism. From *Laboratory Manual for Physical Geology* by Norris Jones. Reproduced by permission of The McGraw-Hill Companies.

■ METAMORPHIC ROCK CLASSES

Metamorphic rocks occur in two broad classes: *foliated* and *nonfoliated.* These are further subdivided on the basis of rock texture and mineralogy.

Foliated Metamorphic Rocks

Foliated rocks are those primarily made up of flat, sheetlike minerals (mainly micas) and generate rock textures characterized by layered or banded sheets of minerals. The mineral orientation occurs as the previous minerals recrystallize within the stress fields of the metamorphism.

The texture and size of the resulting crystals determines the name of the foliated metamorphic rock samples, and these changes are the result of increasing metamorphic grade. *Metamorphic grades* are illustrated in Figure 4.3.

As the temperature and pressure increase, the *precursor rock* slowly changes until some new equilibrium state is reached and a particular metamorphic rock is generated. If the temperatures and pressures are slight, a low metamorphic grade rock called *slate* is formed, composed of very small platy mica crystals. Slate cleaves into wide, smooth, dull-surfaced sheets with a feel and appearance much like the blackboard of your classroom. Slate grains are microscopic, the result of altered clays and mudstones. Slates generally occur in black, green, or reddish tints (Figure 4.4).

If the heat and pressure produce a still higher metamorphic grade, the mica crystals will be relatively larger and produce a *phyllite.* These rocks have crystals that are large enough to be barely visible. The result is a sheen that gives phyllites the appearance of silk or satin cloth in colors of silver, copper, and bronze. Phyllite rock surfaces are often wrinkled, adding to the attractive appearance. Phyllites are intermediate between low and medium metamorphic grade rocks (Figure 4.5).

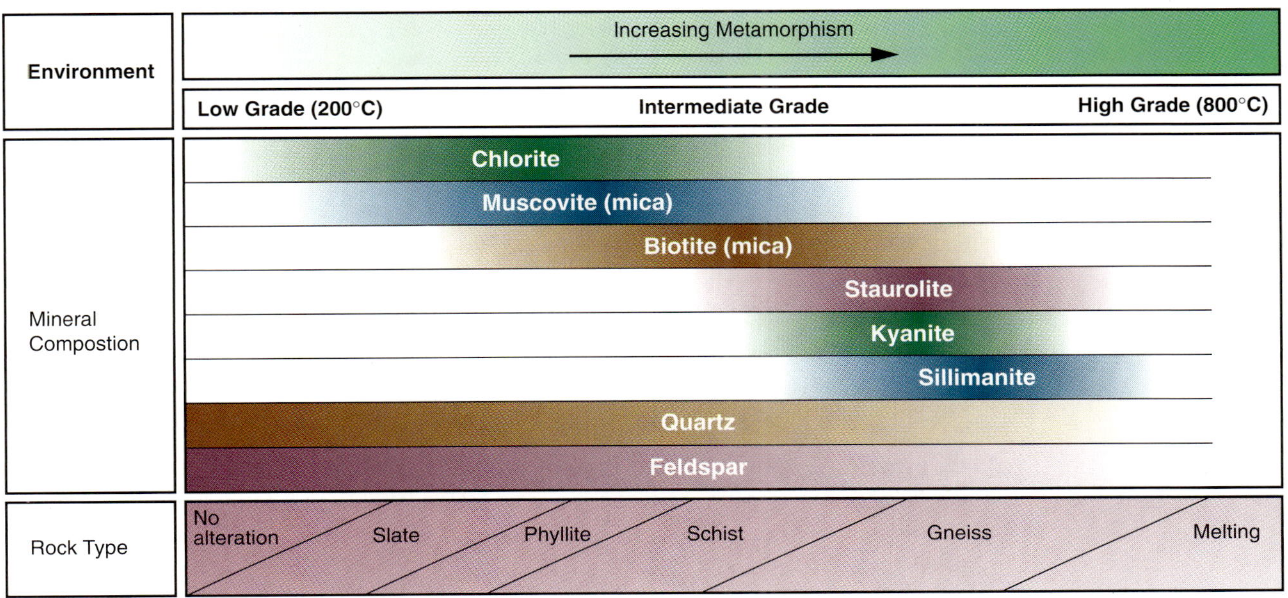

FIGURE 4.3 Metamorphic grades.

FIGURE 4.4 Slate varieties.

FIGURE 4.5 Phyllite.

FIGURE 4.6 Mica schist.

FIGURE 4.7 Schist with garnet porphyroblasts.

Medium-grade metamorphic changes produce larger mica flakes of a size much like fish scales. These are ***schists*** (Figure 4.6) and are formed by the **recrystallization** of lower-grade foliated rocks. The bedding of slates and phyllites is very uniform and even, but schists are irregularly layered.

Schists often contain large crystals of accessory minerals such as pyrite, garnet, and staurolite. Presence of scattered larger crystals, termed ***porphyroblasts,*** will usually tell you that you have a schist sample (Figure 4.7).

High-grade metamorphic conditions cause the micas to recrystallize in part, producing banded rocks where the bands are of different colors. Each layer of these rocks is dominated by a specific suite of minerals, and the rock type is a ***gneiss*** (pronounced "nice") (Figure 4.8).

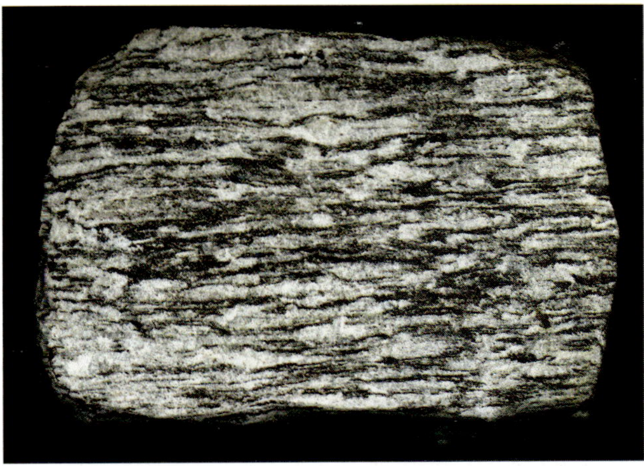

FIGURE 4.8 Gneiss.

FIGURE 4.9 Greenstone.

If foliated rocks are exposed to even higher metamorphic conditions, they cease to be foliated and produce materials such as **migmatites,** rocks that are partially gneissic in nature but may also contain layers and lenses of granite-like materials. These intensively metamorphosed rocks are transitional into igneous rocks.

Nonfoliated Metamorphic Rocks

The second major class of metamorphic rocks are called **nonfoliated.** These rocks lack layering, have few platy minerals, have a crushed grain or other granular appearance, and are often featureless. Mineralogy is generally simple; for example, limestone, which is calcium carbonate, is altered to **marble** with the same chemistry. Another example is sandstone (quartz) changing to **quartzite** with the same mineralogy.

Greenstones (Figure 4.9) are light green, dull, and slightly granular rocks formed by the alteration of basalts, gabbro, or ultramafic rocks (Figure 4.9). **Serpentine** and **soapstone** (talc) are also greenish, soft, and resemble each other. Talc has a soapy feel, and while there is a smoothness to serpentine, there is also a slightly abrasive feel as well. Experience will disclose these differences.

Marble (Figure 4.10) has already been said to be metamorphosed calcite. These granular rocks are light colored and sometimes irregularly streaked or banded ("marbleized" due to layers of impurities). Related rocks are called **skarn,** where the original calcite or dolomite is altered by metasomatism, a process where calcite and quartz are simultaneously metamorphosed. In this type of metasomatism, the result is a mineral called wollastonite ($CaSiO_3$). Skarns are generally more brightly colored than marbles, often being reddish or green in color (Figure 4.11).

Quartzite (Figure 4.12) is a rock formed from quartz sandstone, where the grains of quartz are cemented by quartz. In this rock the broken surfaces show that both the grains AND the cement broke with the fracture through the grains, not around the grains as in a sandstone. These are very tough stones—durable and hard to break.

FIGURE 4.10 Marble.

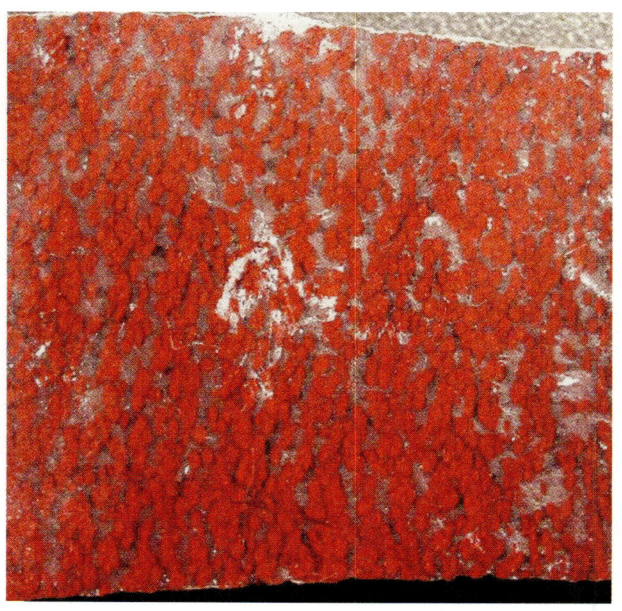

FIGURE 4.11 Skarn.

FIGURE 4.12 Quartzite.

FIGURE 4.13 Amphibolite.

Two dark stone types are **amphibolite** and **hornfels.** In amphibolite, the rock is mainly composed of dark, shiny crystals of amphibole (Figure 4.13). Under a microscope these crystals clearly reveal the same cleavage patterns you learned when we studied minerals in Chapter 2.

If the rocks are exposed to even higher metamorphic conditions, they cease to be foliated and produce materials such as **hornfels,** a nonbanded, granular, dark-colored rock (Figure 4.14).

Anthracite coal is also a metamorphic material, being formed by the metamorphic alteration of **bituminous coals.** These coals are lighter in weight and shinier than their bituminous counterparts (Figure 4.15).

FIGURE 4.14 Hornfels.

FIGURE 4.15 Anthracite coal.

IDENTIFICATION PROCEDURE FOR METAMORPHIC ROCKS

The identification keys in Figures 4.16 and 4.17 will allow you to follow a series of steps that will lead you directly to the correct determination.

If the rock is **foliated,** use Figure 4.16 and check on the grain size to determine the rock name. If the grains are very small, the rock will be a slate; if grains are barely visible and there is a sheen, the rock is a phyllite. If the grains are very evident, being scaly in appearance, or if there are porphyroblasts present, consider a schist, while if the rocks are distinctly color banded, the rock will be a gneiss.

If your sample is **nonfoliated,** first check Figure 4.17 and determine how hard the rock is. Greenstone, serpentine, soapstone, and marble will not scratch glass while amphibolite, quartzite, and hornfels will scratch glass easily. Marble and skarn will effervesce (fizz) when dilute hydrochloric acid is applied.

Grain-size Class and Diameter	Rock Names		Comments
Microscopic, very fine-grained	SLATE		Slaty cleavage well developed
Fine-grained	PHYLLITE		Phyllitic texture well developed; silky, shiny luster
Coarse-grained, macroscopic, mostly micaceous minerals or prismatic crystals; often with porphyroblasts	SCHIST	MUSCOVITE SCHIST CHLORITE SCHIST BIOTITE SCHIST TOURMALINE SCHIST GARNET SCHIST STAUROLITE SCHIST KYANITE SCHIST SILLIMANITE SCHIST AMPHIBOLE SCHIST	Types of schist recognized on the basis of mineral content.
Coarse-grained; mostly nonmicaceous minerals	GNEISS		Well-developed color banding due to alternating layers of different minerals.

Increasing Grade of Metamorphism

FIGURE 4.16 Key to foliated metamorphic rocks.

Metamorphic Rock	Comments
QUARTZITE	Composed of interlocking quartz grains
STRETCHED-PEBBLE CONGLOMERATE	Original pebbles distinguishable, but strongly deformed
GREENSTONE	Composed of epidote and chlorite; green
AMPHIBOLITE	Composed of amphibole and plagioclase; coarse-grained
HORNFELS	Composed of pyroxene and plagioclase; fine-grained
HORNFELS	Composed of quartz and plagioclase; fine-grained
MARBLE	Composed of interlocking calcite or dolomite grains
SKARN	Composed of calcite and added minerals; multicolored
SERPENTINITE	Composed chiefly of serpentine; greens
SOAPSTONE	Composed chiefly of talc; soapy feel
ANTHRACITE COAL	Bright, hard coal; breaks with conchoidal fracture
GRAPHITE	Soft, dark gray, with greasy feel

FIGURE 4.17 Nonfoliated metamorphic rock classification.

■ FURTHER QUESTIONS

1. Although fewer types of metamorphic rocks are used in our everyday existence, can you think of some? Slate blackboards and slate roofs are a couple, and some metamorphic rocks are used for building purposes. Perhaps you have a slate or marble floor or countertop in your home.

2. Use the Internet to look for sites discussing metamorphic rock deposits and determine which minerals are being utilized in industry. What economic importance is there in these deposits?

3. The Internet will also have information of seafloor hydrothermal vents, the so-called "black smokers." These hot water vents are rich in minerals and nutrients producing interesting seafloor deposits. Some investigators believe that such locations are possible sites for the origin of life on earth.

Data Sheets for Metamorphic Rocks

Sample #	Rock Name	Texture	Composition	Other Properties How did you determine this?

Data Sheets for Metamorphic Rocks

Sample #	Rock Name	Texture	Composition	Other Properties How did you determine this?

Data Sheets for Metamorphic Rocks

Sample #	Rock Name	Texture	Composition	Other Properties How did you determine this?

Data Sheets for Metamorphic Rocks

Sample #	Rock Name	Texture	Composition	Other Properties How did you determine this?

Introduction to Maps and Mapping

5

aps have long been used by ancient and modern societies for numerous purposes. One of the oldest surviving maps and the oldest geological map is an ancient Egyptian scroll that detailed the route to gold fields of Wadi Hammamat, a caravan route between the Nile River and the Red Sea. This map was prepared during the reign of the Pharaoh Rameses IV (1151–1145 B.C.E.), drawn and designed by a professional scribe named Amen-Nakhte, son of Ipuy. This surprisingly accurate map even shows the general geology along the route (James A. Harrell and V. Max Brown, 1992. "The oldest surviving topographical map from ancient Egypt" [Turin Papyri 1879, 1899, and 1969], *Journal of the American Research Center in Egypt,* Vol. XXIX, pp. 81–105). (See Figures 5.1 and 5.2).

Ancient maps are often hard to appreciate with a modern eye. The Egyptian goldfield map shows what looks like a speckled snake running between two rows of wedges. Colored wedges depict mountains, the color of those mountain masses record actual changes in rock character and the "snake" represents an ancient rock-paved roadway. The map colors depict six distinct geologic units: Black hills are the Hammamat graywackes; pink hills are the Dokhan volcanic complexes, the Atalla Serpentinite and the Fawaldiir granites; the pink hill with brown lines is the Fawakbir granite with the iron-stained, gold-bearing quartz veins; and the spotting along the road represents the coarse gravel flooring the desert wadis (dry valleys).

This map was certainly not made to represent the geology in the way we would do it today, but it was prepared to show the locations of important quarry sites. Interestingly, the next geological maps we know of were prepared in France during the mid-1700s, nearly three thousand years later!

FIGURE 5.1 Ancient Egyptian map. Brown mountains are graywackes.

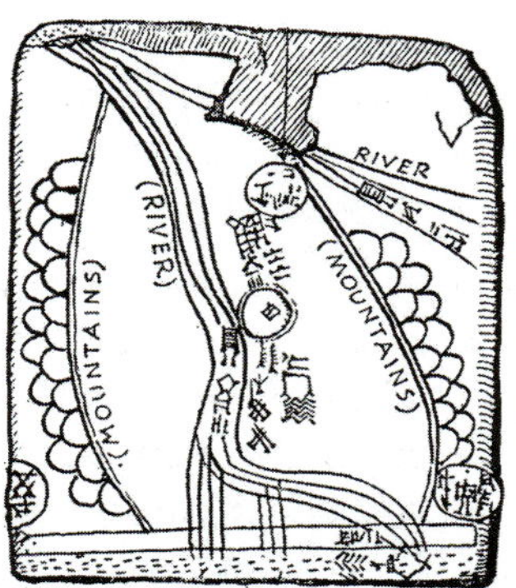

FIGURE 5.2 Ancient Egyptian map. Striped mountain (top) shows granite with iron-stained gold-bearing quartz veins.

Even earlier maps were made in Mesopotamia, the earliest maps in existence apparently being several on clay tablets. One from Ga-Sur, a city north of Babylon, shows a river (Euphrates River?) flowing between two rows of mountains and depicting the locations of several towns (Figure 5.3, redrafted from Erwin Raisz, 1948. *General Cartography.* McGraw-Hill).

Another Babylonian map (Figure 5.4, from David Greenland and Harnl J. De Blij, 1977. *The Earth in Profile.* Canfield Press) locates Babylon, the Tigris and Euphrates Rivers, and several mountain ranges. It places Babylon in the center of the known world—a common feature of the earliest maps. The kingdom of Babylon is shown as

FIGURE 5.3 Babylonian map of the region around the ancient city of Ga-Sur in modern Iraq.

FIGURE 5.4 Babylonian map.

the focus of a disk-shaped world floating in the midst of a world ocean. The same viewpoint was still in use up to the time of the onset of the great westward explorations in the 1500s.

The ancient Babylonians used a number system based on twelves, not tens. This system is directly responsible for the division of circles into 360 degrees, each degree into 60 minutes, and each minute into 60 **seconds of arc** (Erwin Raisz, 1938. *General Cartography.* McGraw-Hill, pg. 5, Reproduced with permission.).

Early maps generally recorded trade routes or boundaries of cities and nations, and others recorded land ownership. One Roman map survives to show the road system of the Roman Empire between Crete and the Nile in Egypt. This elaborate map located towns, named the roadways, listed mileposts, named mountain ranges, and in many ways probably served the same purpose as our modern planometric road maps.

Accuracy and scale were often ignored in the early maps, because it was apparently more important to record the names and general direction of places than to record the details we demand today. A Roman map of the Nile River is ridiculously out of scale—the Nile itself depicted in a nonrealistic manner and the edge of the river studded with

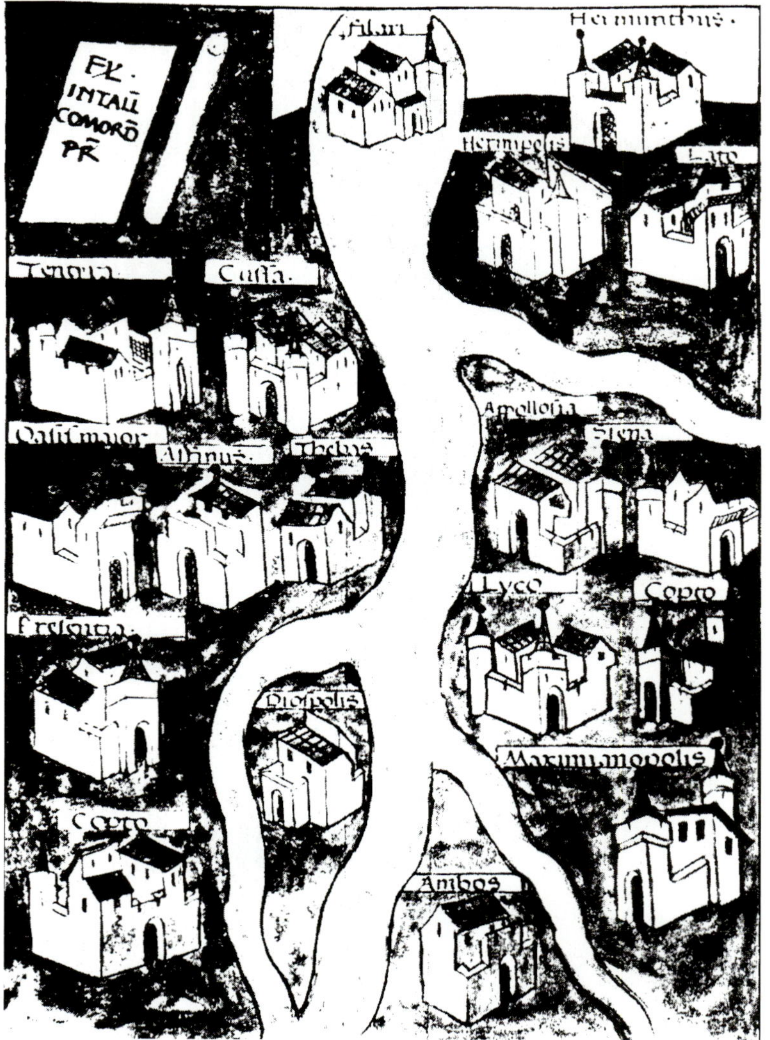

FIGURE 5.5 Roman map of the Nile.

large bulky sketches of multistoried castle-like Roman buildings, a type of construction essentially unknown in Egypt (Figure 5.5, modified from Leo Bagrow, 1966. *History of Cartography*. Cambridge-Harvard Univ. Press).

Maps produced by Greek and Roman authors are sometimes oddly symmetrical, because the authors wanted to show distances and locations in more symbolic than literal terms. The idea of accurate scale, precise spatial location, and distance indications are modern treatments.

Like the Babylonians before them, a famous Roman map depicts Europe, Africa, and Asia as essentially equal in area and arrayed symmetrically around the Mediterranean Sea. The entire map area is circular and again is located in an apparently endless world ocean (Figure 5.6, modified from Erwin Raisz, 1948. *General Cartography*. McGraw-Hill. Reproduced with permission.).

Other cultures had their special mapping tools and methods. An Aztec map shows city locations and some limited geographic features, and marks the roadways with footprints to make their purpose especially clear (Figure 5.7, modified from Erwin Raisz, 1948. *General Cartography*. McGraw-Hill. Reproduced with permission.).

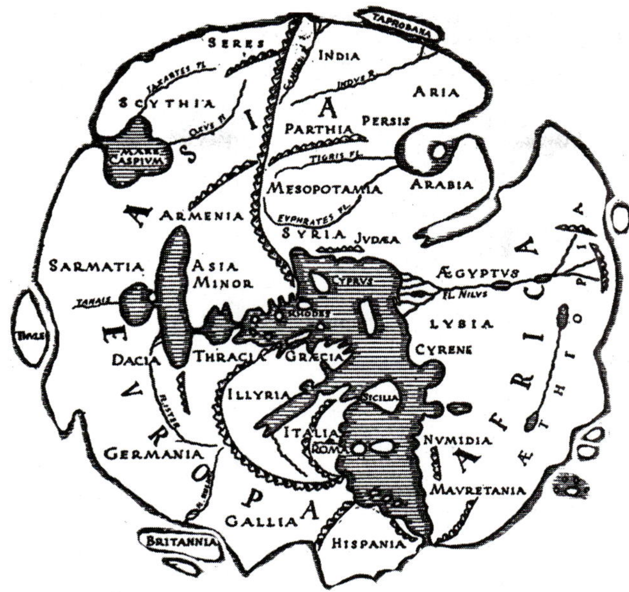

FIGURE 5.6 Roman world map.

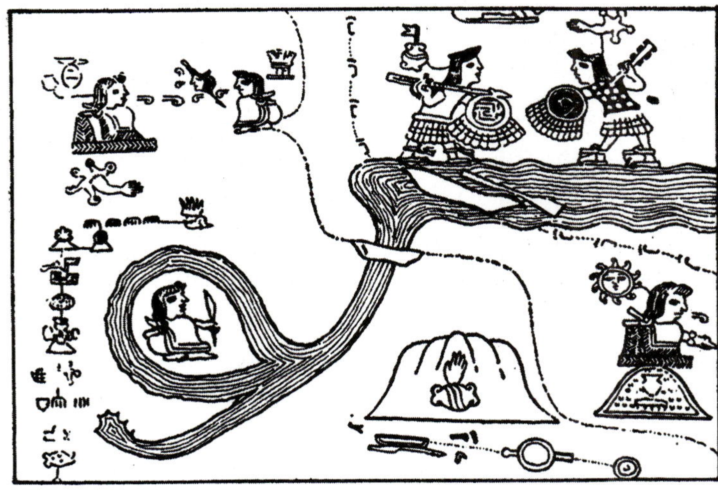

FIGURE 5.7 Aztec map.

Even more curious to Western eyes are the palm frond and shell navigational maps used by Pacific Ocean island cultures. Marshall Islanders created these complex frameworks to help them navigate without a compass as they sailed back and forth across large parts of the Pacific Ocean. Straight sticks served as frameworks and represented the open sea while curved sticks depicted the shapes of the oceanic wave fronts. Cowry shells were tied into the frame to show the location of island groups. (See Figure 5.8.)

The Renaissance and the Age of Exploration brought mapmaking forward, but as late as 1475, maps still showed the world as a circular, flat object in the midst of the great world sea, even though mapmakers knew full well that the world was spherical. Nct until the Western movement of exploration found North and South America did mapmakers truly begin to make maps in a more realistic and accurate manner.

FIGURE 5.8 Polynesian map.

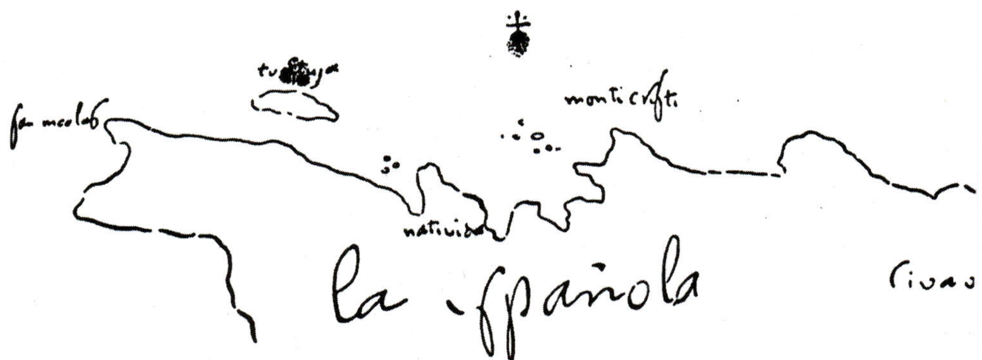

FIGURE 5.9 Columbus's map of parts of the island of Hispañola.

Sailors such as Columbus absolutely had to have accurate maps in frontier areas where even slight mistakes meant that their ship might be wrecked and that they would very likely never see home again. A sketch map by Columbus survives, showing the north shore of the island of Hispañola (Figure 5.9, modified from Leo Bagrow, 1966. *History of Cartography.* Cambridge-Harvard Univ. Press).

PLANOMETRIC MAPS

Today, for everyday use, we mainly use maps to help us find the way between two or more locations, exactly the same need that drove the ancients to prepare their maps. Everyone is familiar with the standard folded road maps that we use to find our way on vacations or to find our way through cities. These are generally referred to as **planometric maps** and record the data on a two-dimensional surface (Figure 5.10).

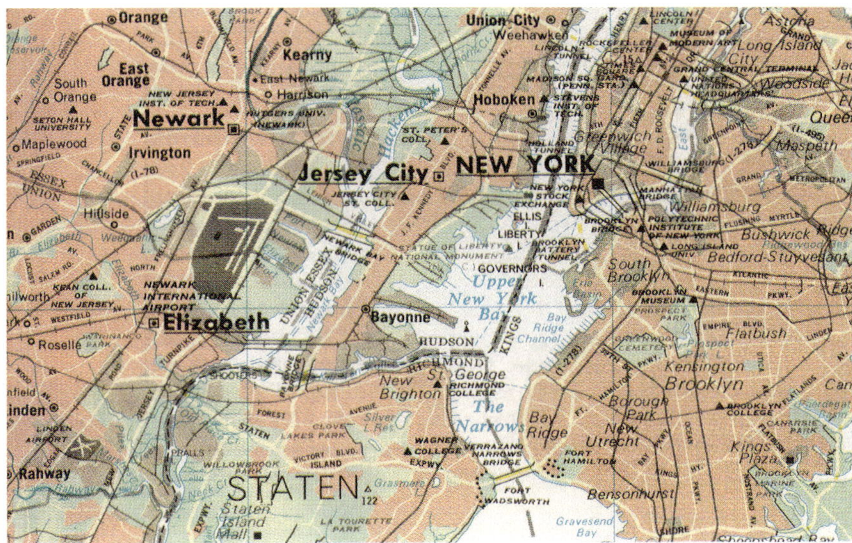

FIGURE 5.10 to 5.12   Planometric map of the New York City area (top), Oblique aerial view of the same area (left, below), and high altitude photo (right, below) From *Goode's World Atlas,* 19th edition, www.randmcnally.com by E. B. Espenshade © by RMC, R. L. Rand McNally & Co., 1998, reprinted with permission.

Compare the detail of a good quality planometric map (Figure 5.10) with a low-altitude oblique photograph of the same area (Figure 5.11) and the same area from a high-altitude view (Figure 5.12).

Although very useful for general work, planometric maps tell nothing about the shape of the ground surface between locations. For instance, a map of Colorado and Utah would show the locations of Denver, Colorado, and Salt Lake City, Utah and allow the user to trace the route between them . . . but it would completely omit the fact that there is a huge mountain range between the two cities.

Other types of planometric maps commonly seen are those used by aviators and mariners. Aviators are generally not too worried about the landforms below but are mainly concerned about straight-line distances and locations of navigation aids, airports, and potential hazards. Their planometric maps record such useful data but mostly ignore the data that make road maps so useful to the driver of an automobile.

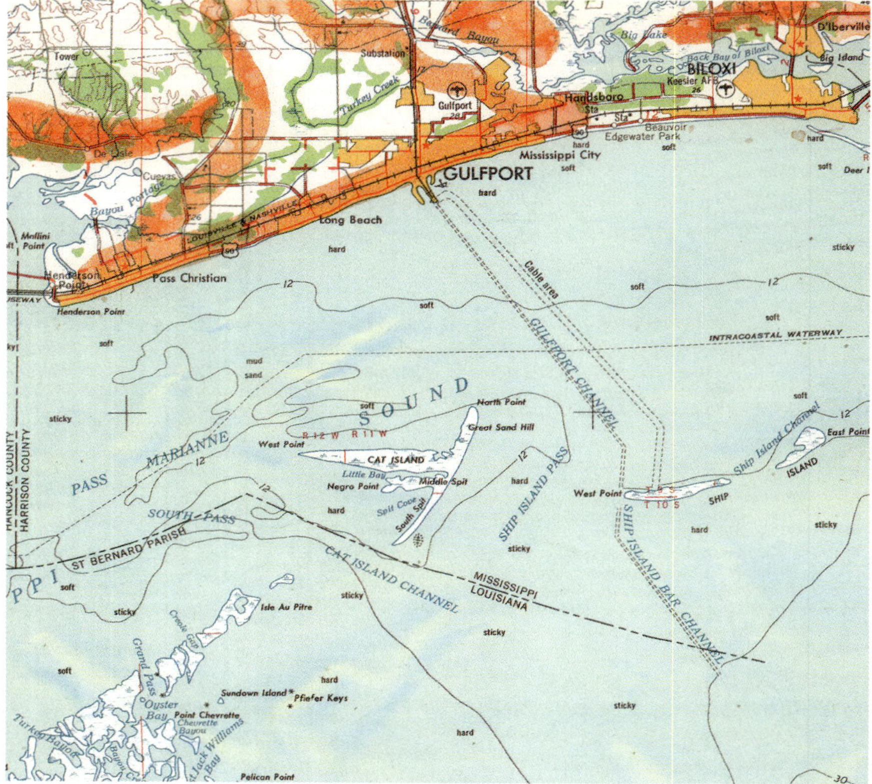

FIGURE 5.13 Marine coastal navigation map, offshore Louisiana. From U.S. Coast and Geodetic Survey.

In the same manner, a person at sea has little use for surface mapping information since, for all practical purposes, the sea is a flat surface (although the sailor certainly does need to know the shape of the seafloor beneath his vessel). A sea captain mainly needs maps or charts to help find the location of ports, headlands, and the location of hazards that lurk beneath the surface of the waters (Figure 5.13).

■ TOPOGRAPHIC MAPS

Three-dimensional mapping produces products generally known as **topographic maps.** These maps combine information from planometric maps with another layer of data that indicates height above or below sea level. These special information lines are termed contour lines. In Chapter 6 we will learn to interpret **contour lines** on topographic maps and even learn to make our own versions.

Topographic maps have many uses in everyday life. Hikers and campers will be able to locate the best, safest, and least strenuous routes by determining how steep a slope might be, where dangerous canyons are located, the best stream crossings, and similar data. Hunters can use topographic maps to locate their prey; fishermen use contour maps of lake bottoms to determine water depth; and sailors employ such maps to help them stay in deep enough water to avoid running aground.

Before you can use **topographic maps** and other special-purpose maps, however, there are a number of map features that you need to understand.

Longitude and Latitude

Understanding that the world was essentially spherical, early geometers and geographers divided the globe into standard geometric subdivisions. The largest of these divisions is the degree and is shown on a map by a number and a special symbol—for example, ninety degrees is written as 90°. Since a circle is divided into 360°, so is the circumference of the earth. Therefore, we see lines running from the North geographic pole to the South geographic pole and crossing the equator. These North–South lines are called lines of *longitude* (Figure 5.14).

Measurement in degrees is too coarse for most surveys, so geographers subdivide each degree into 60 minutes (shown on maps as a number and symbol [′]). Each of the 60 minutes is further subdivided into 60 seconds (shown as a number and symbol [″]). There are further subdivisions possible, but they are seldom used for ordinary surveys. Therefore, 1°= 60′ = 3600″ of arc.

In order to make uniform measurements in **longitude,** a starting point of zero longitude was decreed as being a line running North-South through Greenwich, England. This zero point line of longitude is termed the *prime meridian* (also referred to as the *Greenwich Meridian*).

Longitude is then measured westward toward the United States in degrees of West Longitude and east toward the Orient as degrees of East Longitude. East and west longitude numbers will always equal 180° or less, as the two systems meet on the opposite side of the globe. The meeting place is also called the *International Date Line,* the point where travelers lose a day going West . . . and gain a day going East.

In order to find the location of an exact point on a sphere, you need two intersecting lines. This is accomplished by lines parallel to the equator called lines of *latitude.* These lines of latitude begin at zero degrees at the equator and increase to a maximum of 90 degrees at the north or south geographic poles. These degrees are also subdivided into minutes and seconds. Therefore, latitude measurements are always measured at or less than 90° north or 90° south latitude.

Lines of longitude run from the North Pole to the South Pole, but are measured in degrees east or west of the Prime Meridian, and the number of degrees never exceeds 180°. When you get to 180° east or west longitude, you have reached the opposite side of the globe and the International Date Line.

IN SIGHTS

Lines of latitude are always parallel to the equator and are measured north and south from the equator, and the number of degrees never exceeds 90°. When you reach 90° north latitude, you are at the North Pole, and when you reach 90° south latitude, you are at the South Pole.

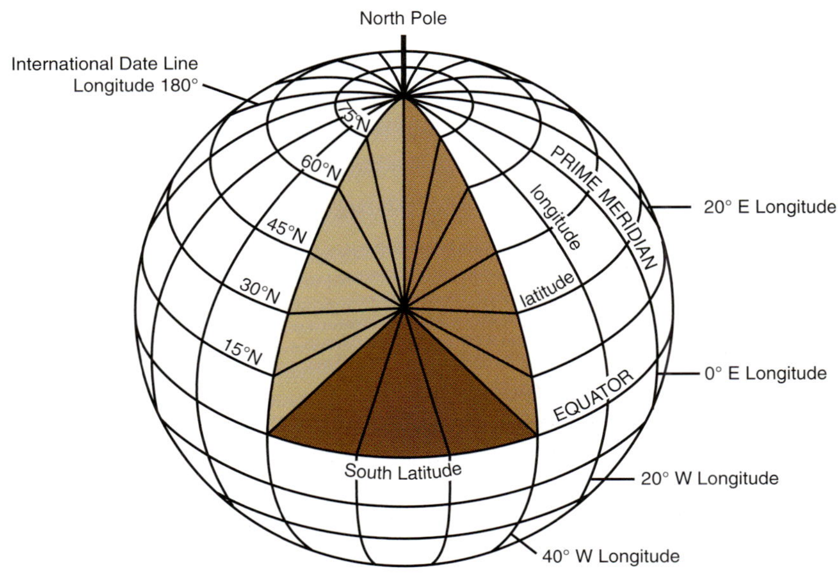

FIGURE 5.14 Longitude and latitude.

IN SIGHTS

You absolutely must state whether longitude is east or west and whether latitude is north or south. Your instructor can demonstrate that without the compass direction information, your longitude and latitude will find FOUR different locations—not one unique site.

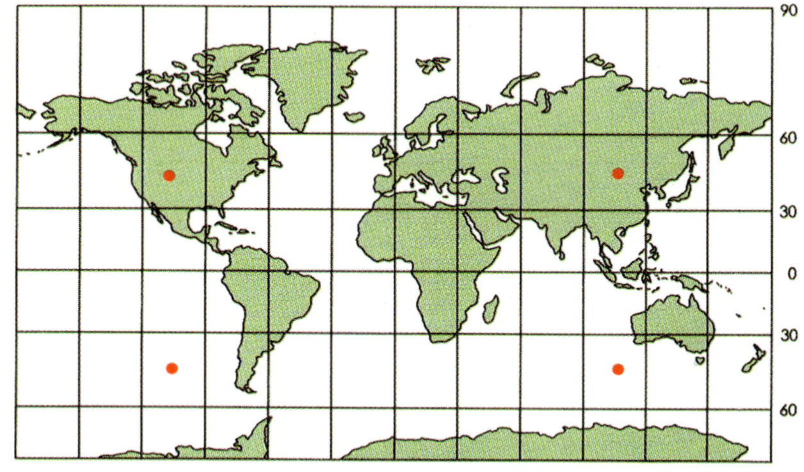

FIGURE 5.15 Possible landing sites. From *Goode's World Atlas,* 19th edition by E. B. Espenshade © by RMC, R. L. 00-S-38, www.randmcnally.com.

Imagine that you were telling a pilot to fly to longitude 105° and latitude 40°—the approximate location of Denver, Colorado—but you didn't give the pilot any compass directions. If the pilot was unskilled enough, he or she might go to longitude 105° east and latitude 40° north and end up in the Ordos Desert of China, or perhaps to longitude 105° east and 40° south to crash land in the Indian Ocean hundreds of miles southwest of Australia, or to longitude 105° west and latitude 40° south to make a similar crash landing in the Pacific Ocean west of Chile. The only coordinates that would bring the pilot safely to Denver would be longitude 105° west and latitude 40° north (Figure 5.15).

MAP PROJECTIONS AND THE PROBLEM OF DISTORTION

Although ancient Greek and Roman mapmakers were well aware of the fact that the earth was round, they struggled mightily to depict this round surface on a flat sheet of paper. No matter how you do try to transfer the rounded shape of the planet to the flat surface of a sheet of paper, the map surfaces will distort the shapes of countries and seas.

Today there are many different projections—ways to depict the spherical earth on a flat paper. Each of these projections strives to give the most accurate map for a given purpose. Figure 5.16 (*Goode's World Atlas,* 19th ed. Rand McNally & Co., 1998, printed with permission) illustrates several of the more popular projections seen in books and magazines today.

Keep an eye on North America as you scan down the five different projections. Our continent is distorted in all of these; however, a map similar to the upper one (Miller Circular Projection), with the map center located in the middle of the United States, would show our country much more accurately but then it distorts other areas. No map is a perfect depiction of the entire planetary surface.

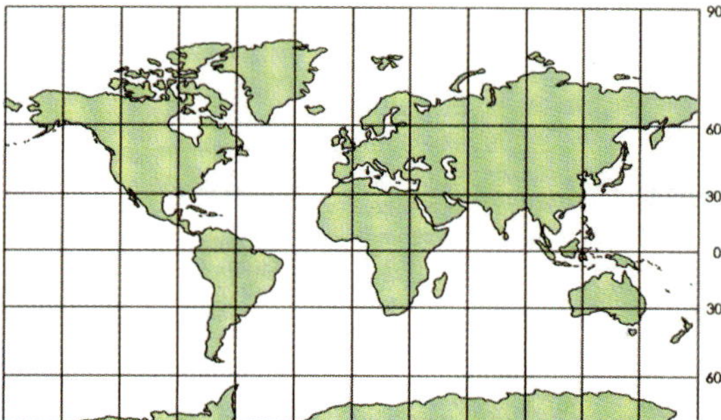

Miller Cylindrical Projection.

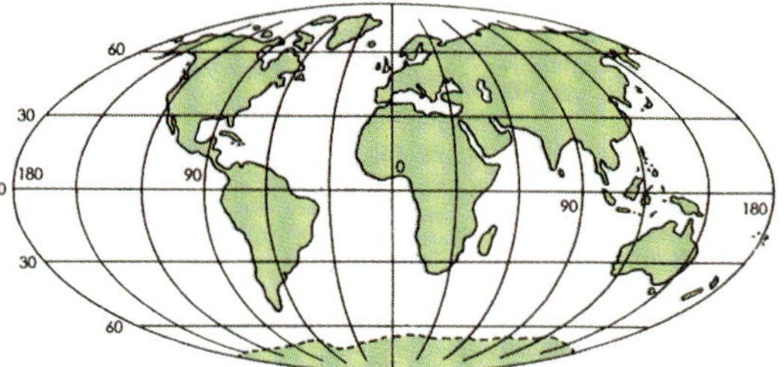

Mollweide Homolographic Projection.

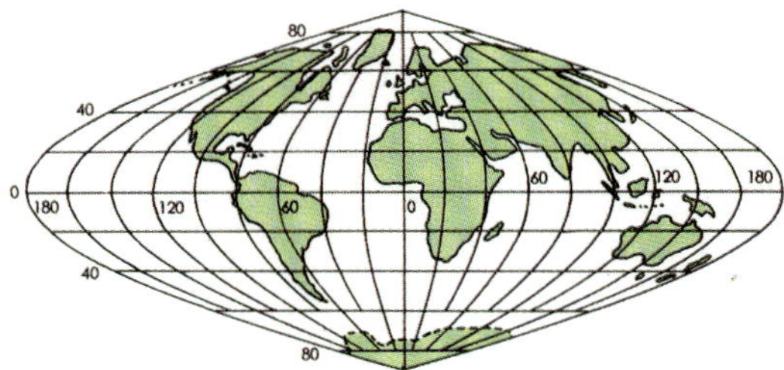

Sinusoidal Projection.

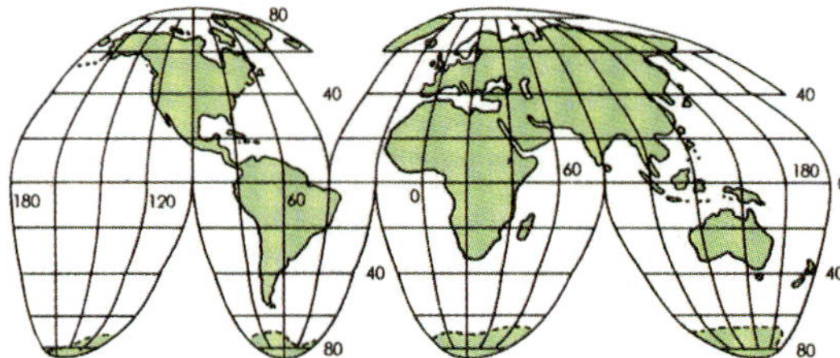

Goode's Interrupted Homolosine Projection.

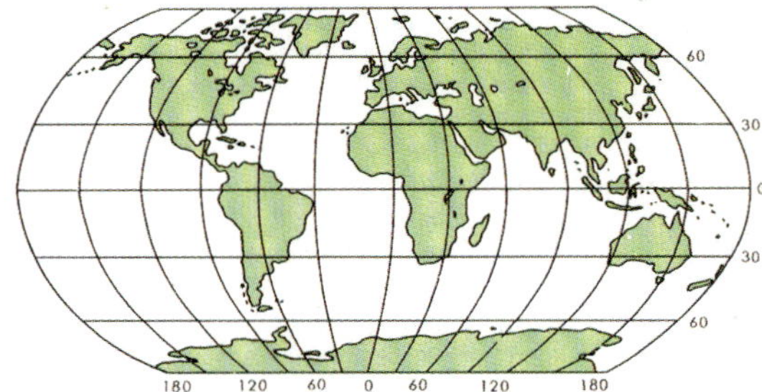

Robinson Projection.

FIGURE 5.16 Common map projections. From *Goode's World Atlas,* 19th edition by E. B. Espenshade © by RMC, R. L. 00-S-38, www.randmcnally.com.

■ FINDING YOUR WAY AROUND THE MAP

Map Scales

Maps of various sizes need to have scales to allow us to see how they relate to the actual distances measured on the ground. One common type is the *ratio scale* where a measurement on the map is equal to some multiple of that measurement on the earth's surface, such as 1:24,000 (also 1/24,000)—one inch on the map equals 24,000 inches on the actual land surface. In this system any convenient measurement can be used.

A nonsensical ratio scale example might be to map using "nose lengths" where one unit on the map as long as your nose is 62,500 "nose lengths" measured on the ground. The unit chosen is immaterial as long as the right ratio is stated on the map.

Ratio scales are often expressed as 1:24,000; 1:62,500; 1:250,000; and 1:1,000,000.

The third type of scale is a *bar scale* printed across the bottom of many maps. Here, a graphical bar resembles a ruler and gives varying units of map measurement. Commonly, these bar scales include feet, miles, and kilometers.

Reference Points

Topographic maps have standardized points of reference located around the sheet to help you understand the map and to locate yourself. (Figure 5.17 is an excellent source of information to help you locate yourself on a standard U.S. Geological Survey quadrangle map. These hints include the following:

- **Longitude and latitude** markings are located along each edge of the map. Check for numbers stated in units like 25° 15′ 30″ where the " ° " equals degrees; the " ′ " equals *minutes of arc*; and the " ″ " equals seconds of arc.

- **Township and range lines** are located at the map edges. Look for units expressed as "T15N" (Township 15 North) and "R27E" (Range 27 East). Thin red lines on the map surface indicate boundaries of sections (units one mile square), and each section will also have a red number in its center to locate you within the township.

- Map direction indicators are located near the bottom of the map sheet and include arrows or other indicators for: True North, Magnetic North, and Grid North.

- Map bearings are commonly expressed in quadrants—for example: North, Northeast, East, Southeast, South, Southwest, West, and Northwest. You will soon learn to use these quadrants to subdivide and label a section into quarters.

- Next-door neighbors. Suppose you were planning to travel across a mapped area and your destination was at a point beyond the present map sheet. Labels along each of the four map borders and at each of the four corners tell you which map sheets are adjacent to those points.

- Name of the map: You will need the official name of the map when you order a copy of a needed map.

- The date the map was created is also important, especially in developed areas where location names, roads, and open areas are in constant change. Look near the base of the map sheet.

- Who made it? Maps are created by a number of agencies and private companies. You will need this information to order copies.

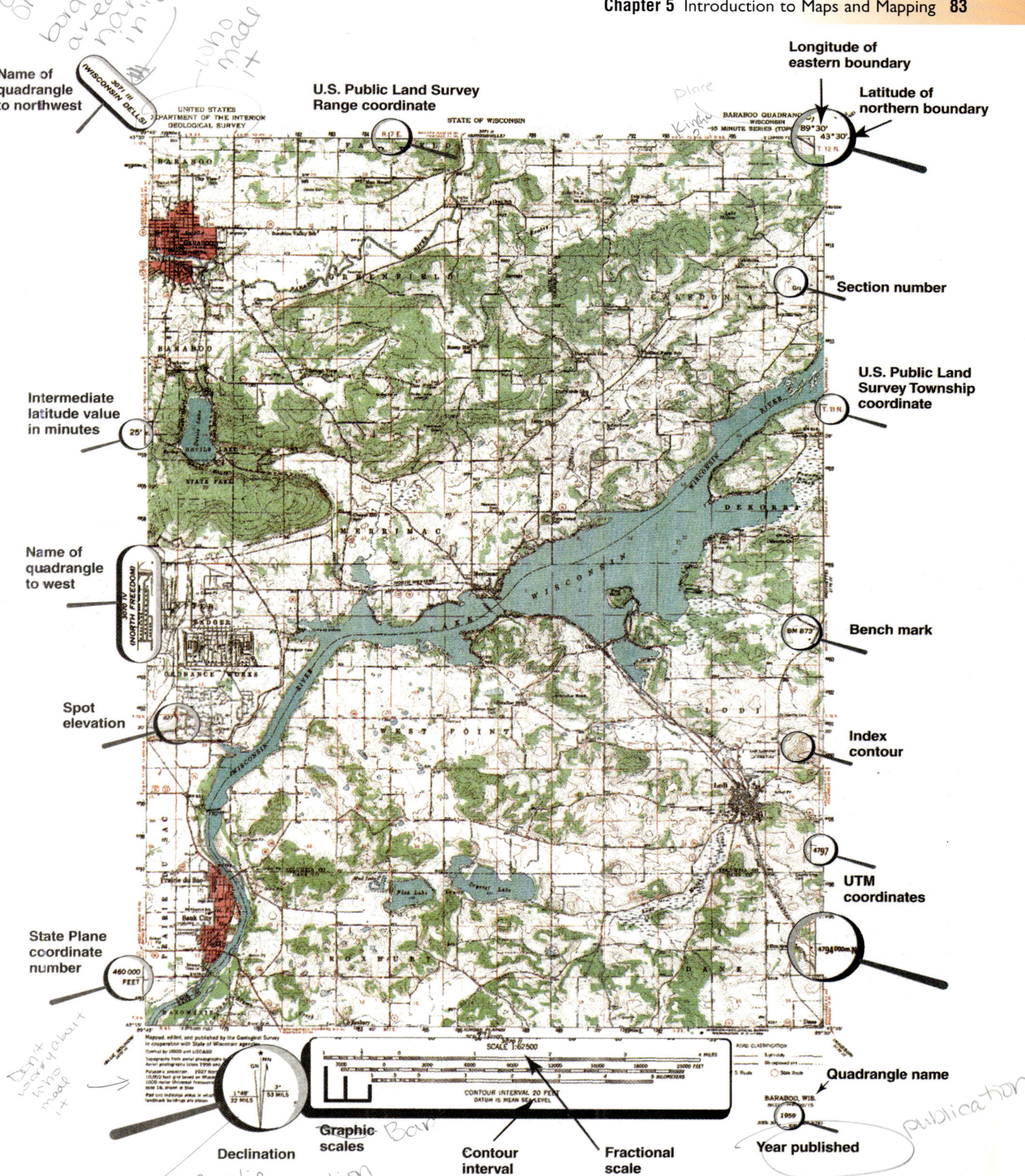

FIGURE 5.17 Index information for the Baraboo, Wisconsin 15-minute quadrangle map. From *Laboratory Manual for Physical Geology,* 2nd ed. by Norris W. Jones. Copyright © 1998 by McGraw-Hill. Reprinted by permission.

- Map symbols are more or less standardized, but if unusual ones are used on your map, they will commonly appear on the sheet with an explanatory label.

- Map colors indicate general features. Greens are used for vegetated areas, blue for water, brown for general land surfaces and hilly areas, and purple (or reds and pinks) commonly are used to show areas near cities that have undergone much recent development and changes since the previous edition of the map.

■ QUADRANGLE MAPS

Longitude and latitude are the basis for what are called quadrangle maps. Quadrangle maps in North America are always taller than they are wide, but, wherever they are encountered, the top and bottom of the maps are latitude lines while the right and left margins of the map are lines of longitude. Standard maps are often seen in scales referred to as 15-minute and 7.5-minute quadrangle maps.

These numbers tell us that on a 15-minute map, the top and bottom margins are 15′ of latitude apart, and the right and left margins are 15′ of longitude apart. On a 7.5-minute quadrangle map, the right and left margins are 7.5 minutes of longitude apart, and the top and bottom margins are 7.5′ of latitude apart.

Because of this, quadrangle maps are nearly square at the equator but very narrow as they near the poles, though the numbers of degrees across the top and sides remain the same. This is to say that, even though the lines of longitude are getting closer together at the poles, they remain the same distance apart in **degrees of arc.**

You must also be aware that the magnetic compass does NOT generally point to the **north geographic pole,** but to the **north magnetic pole** (Figure 5.19). The magnetic pole lies northwest of Hudson Bay in Canada, hundreds of miles from the true north pole.

Further, the magnetic poles slowly drift across the landscape, requiring that maps be periodically updated to avoid compass errors.

If you look at the bottom of your map, you will see a symbol made of two or three arrows. One arrow points directly to the top of the map, indicating true geographic north, while the other arrows have different orientations. One arrow points toward the magnetic pole location at the time the map was compiled and is labeled MN for Magnetic North. If present, the other arrow points North on a special worldwide map grid that is used to minimize map distortion, and this arrow is labeled GN for Grid North.

A 15-minute quadrangle can be subdivided into four 7.5-minute quadrangles. Therefore, a 15′ map covers a larger area than a 7.5′ map but does so with less detail (Figure 5.18).

Each of the degrees, minutes, and seconds is spaced equally and evenly around the globe. However, lines of longitude come together at the poles, becoming ever closer as they extend north or south. Therefore, the distance between two longitude lines varies as you leave the equator. At the north or south geographic poles all longitude lines come together at a single point.

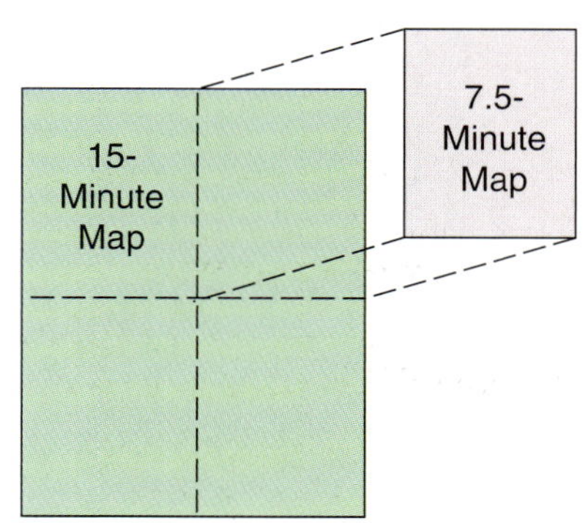

FIGURE 5.18 15′ and 7.5′ map scale example.

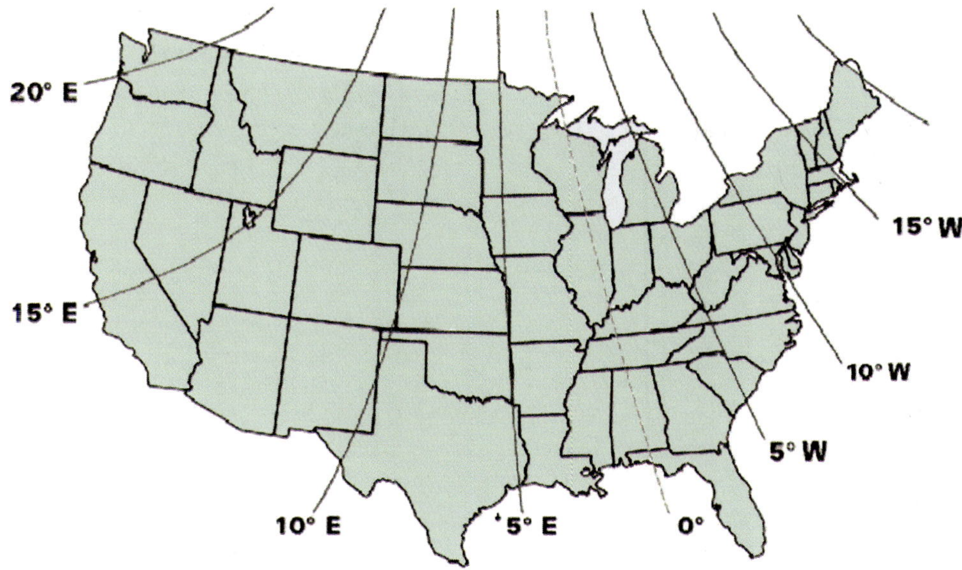

FIGURE 5.19 Magnetic lines of declination of North America.

MAGNETIC DECLINATION

Check these on some maps in your laboratory.

What is the magnetic declination on a map of your city?

What is the magnetic declination of Miami, Florida?

What is the magnetic declination of Seattle, Washington?

What is the magnetic declination of Pensacola, Florida?

What is the magnetic declination of northern Maine?

The difference between the true geographic north arrow, and the MN or GN arrows in degrees is the **magnetic declination.** (See Figure 5.19.) This number must be either added or subtracted (depending whether you are east or west of the magnetic pole) in order to allow you to obtain a true north with your compass (see Chapter 8).

MAPPING SYSTEMS

Longitude and latitude were and are more than accurate enough for long-distance navigation, such as travel by sea or air travel, but this system is difficult to use with the fine scale work needed for ordinary land surveying.

In the early days of North American surveying, a system known as **"metes and bounds"** was employed. George Washington, who worked for a time as a surveyor, used this system. Metes and bounds mapping depends on locating property in relationship to locally understood landmarks such as houses, streams, trees, stone walls, etc. As long as everyone using the system can find all these landmarks, the system works. Once any one of these landmarks is lost, the system becomes chaotic.

This example, taken from an actual 1797 land sale concluded in Pennsylvania, clearly illustrates the difficulty (courtesy of Dr. Carl Withner, personal communication):

> "Now this witnesseth that I, John Zell, for the sum of five shillings HAVE granted, bargained, sold, assigned and set over all that tract or parcel of land situated in Donegal Township, Lancaster County, in the State of Pennsylvania containing fifty-two and one quarter acres and bounded by the following lines and courses to wit.
>
> BEGINNING at a white oak in a line of Randall McClure's land and thence extending by the land of Phillip Hefs, north forty-eight degrees east seventy perches to a hickory grub, thence by the land of George Haugenberger the seven courses and distance following, viz., north sixty-six degrees and a half degree west twenty perches and a half league, thence south fifty degrees west forty-four perches to a post.
>
> West southwest thirty perches to a post. North thirty-six degrees west thirty-five perches to a post, thence by the land of Randolph McClure. South forty-nine degrees west fourteen perches and a half to a chestnut oak, thence by land late of George Bumbaugh deceased, the three following courses, viz., south sixteen degrees east twenty perches, thence to a chestnut tree south seventy-eight degrees west twelve perches to a chestnut oak, south nine degrees west ten perches and three-quarters to a post; thence by land of the said Randle McClure south twenty degrees east seventy-nine perches and a half to a white oak and south fifty-seven degrees east sixty-two perches and a half to the place of beginning containing as aforesaid fifty-two acres and one quarter."

I defy you to easily locate that land today. Those posts and trees that were such important landmarks in 1797 are surely gone today. Plus, does anyone remember where "George Bumbaugh deceased" once lived? And what in the world is a hickory grub?

Worse yet are the units of measure used: In 1797 there were three different measurements for a 'perch', English, French and American. Plus there seems to be no current agreement today on what a 'course' might have been in 1797. And there were five different 'leagues' in use in America at that time . . . two were English, and there were Celtic, Spanish and French leagues as well.

And what about those compass directions. We can be pretty sure the compasses were not corrected for magnetic declination nor were their instruments of the quality we demand as a standard today.

If all these other problems were not enough, people were rather indifferent spellers at that time, and John Zell spells his own name, in his own hand, three different ways in this one document. Good luck in following that survey forward through the land record books to the modern era.

In fact, Dr. Carl Withner, the direct descendant of John Zell, who provided this survey, has searched for decades trying to locate the old family lands . . . but they still remain lost.

Texas is also a good case in point, because when the original surveys were completed, Texas was a part of Spain, then later became part of the country of Mexico. Land surveys were isolated—their base lines depending on the orientation of houses—and each survey was uniquely oriented. Later, when Texas first became an independent nation and then part of the United States, state-sponsored surveys were begun and abandoned a number of times, leaving Texas a patchwork of unrelated databases.

Railroads later created their own base lines as did school districts, private companies, and individual landholders. As a result, today there is no single unified survey for Texas, and nearly all the older surveys were done by metes and bounds. Consequently, selling a piece of agricultural property in Texas becomes a very serious and difficult problem as seller and buyer attempt to prove the exact size and location of the land parcel.

Problems raised by **metes and bounds** mapping were recognized very early, and Thomas Jefferson proposed a mapping system that worked well then and is still generally used today in North America. His system is referred to by two names: **Township and Range Mapping;** and the ***U.S. Public Land Survey System (PLS Surveys).*** One researcher said that the beauty of Jefferson's system was that a farmer who understood simple arithmetic—and who had a long rope— could become his own surveyor as he attempted to define property lines on the American frontier.

The PLS system relies on a carefully surveyed north-south line called the ***principal meridian*** and an east-west line crossing it at right angles referred to as the base line. All other parts of the survey refer back to these two reference lines. To keep the numbered smaller units from becoming hard to work with, states and territories were often separately surveyed with their own principal meridian and base line (Figure 5.20).

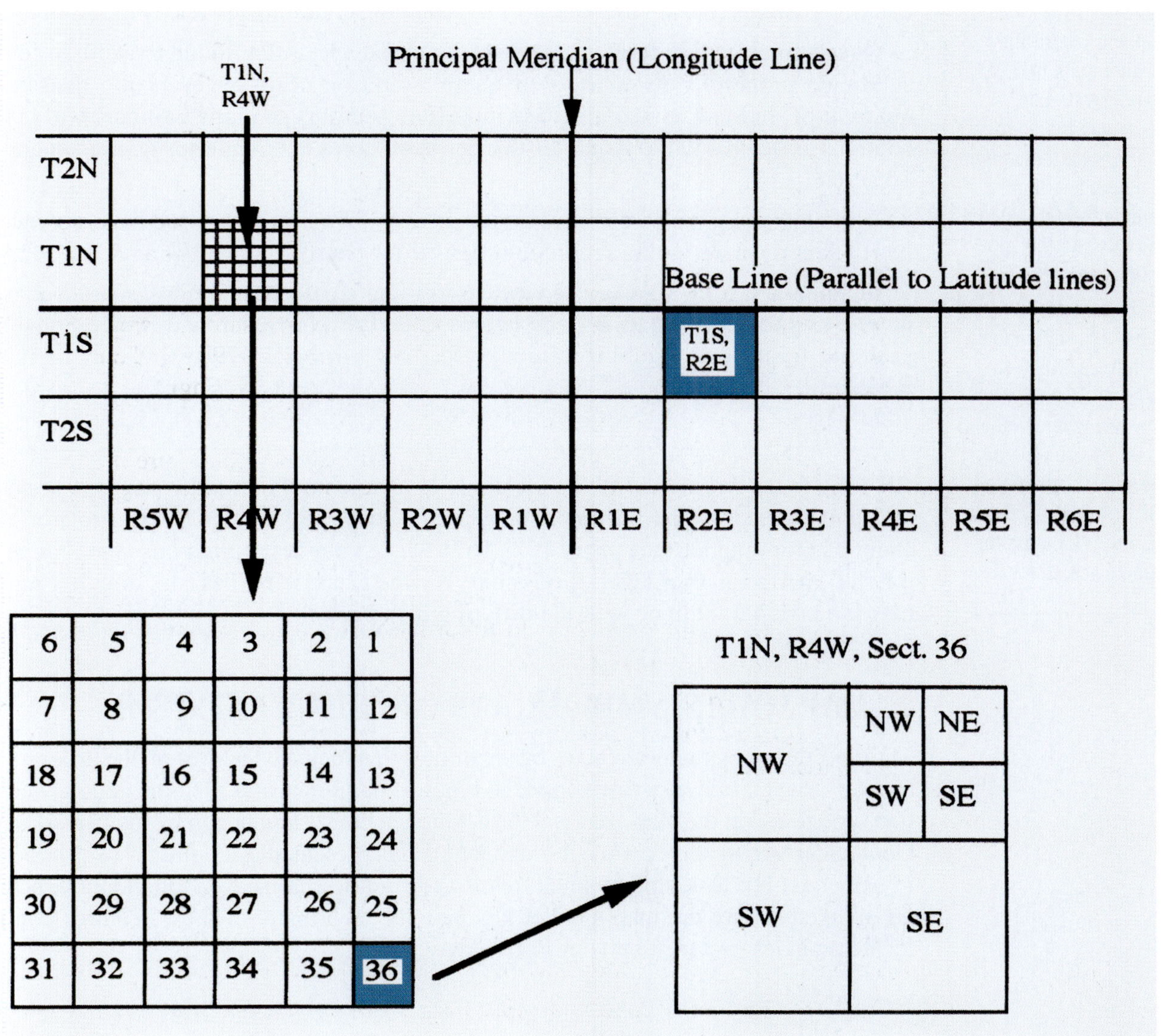

FIGURE 5.20 Public land survey system units.

Due to mapping errors and the fact that lines of longitude come together as they are projected north, some townships will have fewer than 36 sections.

CAUTION

It is absolutely vital that you record the smaller units first. To demonstrate this, determine the quarter-quarter-quarter location of a spot and then reverse the sequence and relocate the spot on the map.

For example, try this. Plot a site within some section as "NE 1/4, NW 1/4, SE 1/4" . . . now see where you end up when you reverse the sequence and plot "SE 1/4, NW 1/4, NE 1/4." Not quite the same place, is it?

Once these regional control lines were set, blocks of land six miles on a side were surveyed, defined by north and south lines called **township lines (also often referred to as tier lines),** while lines arrayed east and west of the principal meridian were termed **range lines.** To determine the location of one of these blocks, simply count the number of range lines east or west from the principal meridian and the number of township lines north or south from the base line. For purposes of identification, the 6-mile-by-6-mile square you have located is also known as a **township.**

Each township ideally contains 36 square miles (6 miles by 6 miles), and each of these one square mile blocks is called a "**section.**" The sections in a township are numbered from the upper right-hand corner across, then down and back from one to thirty-six.

When the great period of western expansion was under way, it was found that one man and his family could easily farm only a quarter of a square mile, and, therefore, the sections were divided into quarters. Each quarter is identified by its compass orientations, that is, north-east, south-east, south-west, and north-west (Figure 5.20).

Once homesteading farmers obtained their land, they often found it useful to further subdivide property into quarters of quarters and on occasion even into quarters of quarters of quarters. Since a square mile (one section) contains 640 acres, a quarter contains 160 acres while a quarter-quarter is 40 acres and a quarter-quarter-quarter is 10 acres.

Although these measurements sound odd to city dwellers, they have served the North American agricultural community well for nearly 200 years.

In summary, a PLS survey is written with the smallest unit first (quarter-quarter-quarter or quarter-quarter) and the largest units last (Township and Range lines). For example, a location might be defined as the "SE 1/4, NE 1/4, NW 1/4, sec. 18, T46S, R14N."

REVIEW: Determine the PLSS locations of a series of landmarks on maps provided by your instructor. The accompanying partial maps in Chapter 8 will provide good examples for practice (see pages 132–167). Make your determinations complete to the quarter-quarter-quarter level, if possible. Try the exercise on both 15′ and 7.5′ minute maps to become familiar with the differing map scales.

■ MAP SCALES, SYMBOLS, AND COORDINATE SYSTEMS

Maps produced by many agencies bear symbols that identify features on the ground such as roads, rivers, railways, towns, special types of buildings, and even the general vegetation type found in the map area. In North America the system begun by the United States Geological Survey has become the standard and is presented as Figure 5.21 (also located in Appendix B). Note that in small towns and rural locations even individual structures are marked. Study the map symbol key and see how many of those features you can locate on maps provided by your instructor.

Topographic Map Symbols

BOUNDARIES

National

State or territorial

County or equivalent

Civil township or equivalent

Incorporated-city or equivalent

Park, reservation, or monument

Small park

LAND SURVEY SYSTEMS

U.S. Public Land Survey System:

Township or range line

Location doubtful

Section line

Location doubtful

Found section corner; found closing corner

Witness corner; meander corner

Other land surveys:

Township or range line

Section line

Land grant or mining claim; monument

Fence line

ROADS AND RELATED FEATURES

Primary highway

Secondary highway

Light duty road

Unimproved road

Trail

Dual highway

Dual highway with median strip

Road under construction

Underpass; overpass

Bridge

Drawbridge

Tunnel

BUILDINGS AND RELATED FEATURES

Dwelling or place of employment: small; large

School; church

Barn, warehouse, etc.: small; large

House omission tint

Racetrack

Airport

Landing strip

Well (other than water); windmill

Water tank: small; large

Other tank: small; large

Covered reservoir

Gaging station

Landmark object

Campground; picnic area

Cemetery: small; large

RAILROADS AND RELATED FEATURES

Standard gauge single track; station

Standard gauge multiple track

Abandoned

Under construction

Narrow gauge single track

Narrow gauge multiple track

Railroad in street

Juxtaposition

Roundhouse and turntable

TRANSMISSION LINES AND PIPELINES

Power transmission line: pole; tower

Telephone or telegraph line

Aboveground oil or gas pipeline

Underground oil or gas pipeline

CONTOURS

Topographic:

Intermediate

Index

Supplementary

Depression

Cut; fill

Bathymetric:

Intermediate

Index

Primary

Index Primary

Supplementary

MINES AND CAVES

Quarry or open pit mine

Gravel, sand, clay, or borrow pit

Mine tunnel or cave entrance

Prospect; mine shaft

Mine dump

Tailings

SURFACE FEATURES

Levee

Sand or mud area, dunes, or shifting sand

Intricate surface area

Gravel beach or glacial moraine

Tailings pond

VEGETATION

Woods

Scrub

Orchard

Vineyard

Mangrove

COASTAL FEATURES

Foreshore flat

Rock or coral reef

Rock bare or awash

Group of rocks bare or awash

Exposed wreck

Depth curve; sounding

Breakwater, pier, jetty, or wharf

Seawall

BATHYMETRIC FEATURES

Area exposed at mean low tide; sounding datum

Channel

Offshore oil or gas: well; platform

Sunken rock

RIVERS, LAKES, AND CANALS

Intermittent stream

Intermittent river

Disappearing stream

Perennial stream

Perennial river

Small falls; small rapids

Large falls; large rapids

Masonry dam

Dam with lock

Dam carrying road

Intermittent lake or pond

Dry lake

Narrow wash

Wide wash

Canal, flume, or aqueduct with lock

Elevated aqueduct, flume, or conduit

Aqueduct tunnel

Water well; spring or seep

GLACIERS AND PERMANENT SNOWFIELDS

Contours and limits

Form lines

SUBMERGED AREAS AND BOGS

Marsh or swamp

Submerged marsh or swamp

Wooded marsh or swamp

Submerged wooded marsh or swamp

Rice field

Land subject to inundation

FIGURE 5.21 Standard U.S. Geological Survey map symbols.

■ MAP COORDINATE SYSTEMS

There are several very different means of locating a spot on maps produced by the United States Geological Survey. The longitude and latitude surveys are indicated on each of the corners of the map in degrees, minutes, and even seconds where appropriate. Check each of the four corners of your map and determine which units are longitude and which are latitude.

The second system is the PLSS system of townships and ranges. Check the borders for statements like T2N or R3W. These will define the sides of the township block. Careful examination of the map will disclose squares outlined in red with numbers in their centers. These squares are sections and are numbered in order back and forth across the township.

The third widely used system is referred to as the UTM system (UTM is an acronym for "universal transverse mercator"). The UTM system employs a rectangular grid that is independent of longitude and latitude and therefore avoids some of the mapping distortions common with the problems of mapping a spherical Earth on a flat sheet of paper.

The UTM system uses a rectangular grid based on a series of 60 north-south zones numbered in order from Zone 01 (between 180° and 174° west longitude) on the left grid margin to Zone 60 (between 174° and 180° east longitude) on the right margin. The north-south lines of the UTM grid are therefore not parallel to longitudinal lines, but instead diverge from longitudinal lines toward the north. The location of a point is described by locating an easting coordinate (a spot measured in meters from west toward the east) and a northing location (measured in meters away from the equator).

You can see the UTM markings along the sides of the quadrangle maps where the locations are printed in special formula format, such as: a northing location $^{44}56^{000}$, which states that the location is 4,456,000 meters north of the equator.

This system is widely used in military mapping, but is not generally employed in civilian work. The results are highly accurate and logical, but we will not use them in this manual.

s. State the verbal scale in miles

—in kilometers

t. On the map one inch equals:

—How many feet on the ground?

—How many miles?

—How many kilometers?

u. On the quadrangle map before you, if the distance right-to-left across the map and if the distance up and down the map are the same in minutes or degrees of latitude and longitude, why is the map rectangular and not square?

Think of the result as a formula with three parts:

- Part One = quarter-quarter-quarter (smallest unit to the left, largest unit to the right)
- Part Two = Section Number
- Part Three = Township (Tier) Number and Range Number

Therefore, when all the parts are assembled in correct order:

Part One + Part Two + Part Three = Correct Location

4. Using the formula in the Insights box on this page, determine the Township and Range (PLSS) coordinates for the house (open rectangle) on the map below. Determine the location to the nearest quarter-quarter-quarter, and remember to include all three parts of the location formula.

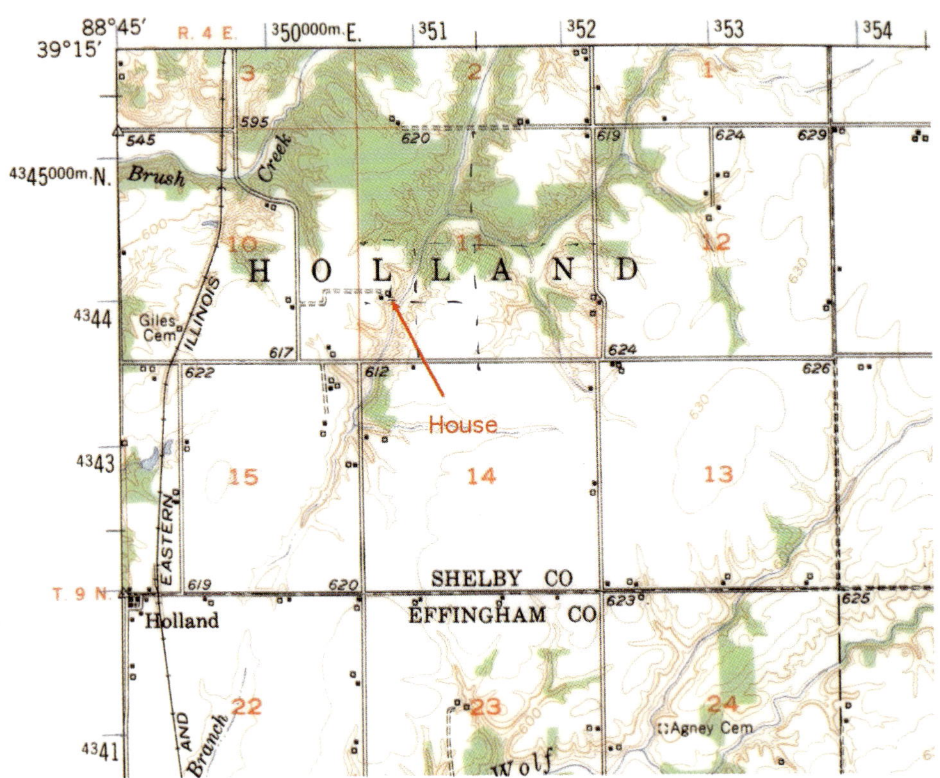

5. Using the image below, determine the quarter-quarter-quarter location for the circle marked 'Comm Tower.' Find the same data for the island in the upper left corner. Remember, you do not have the Township and Range numbers in this example.

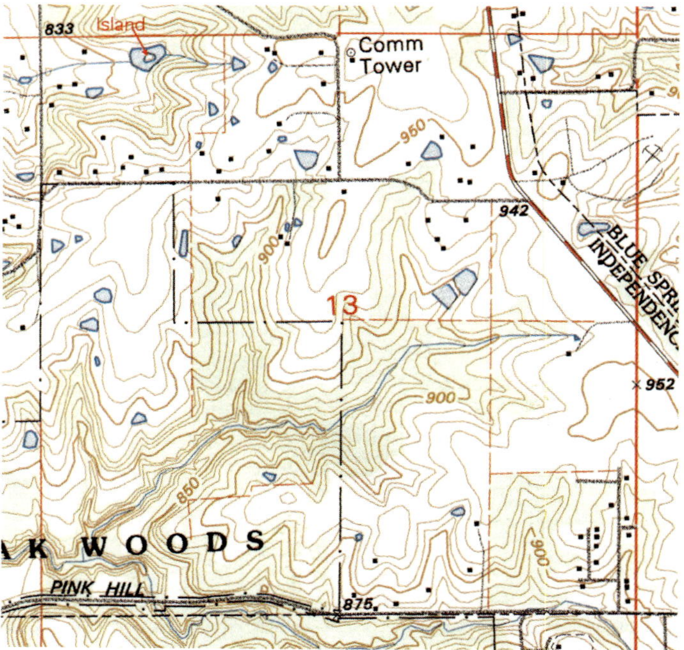

6. In the image below, determine the quarter-quarter-quarter location for the circle marked 'Quarry.' Remember, you do not have the Township and Range numbers in this example.

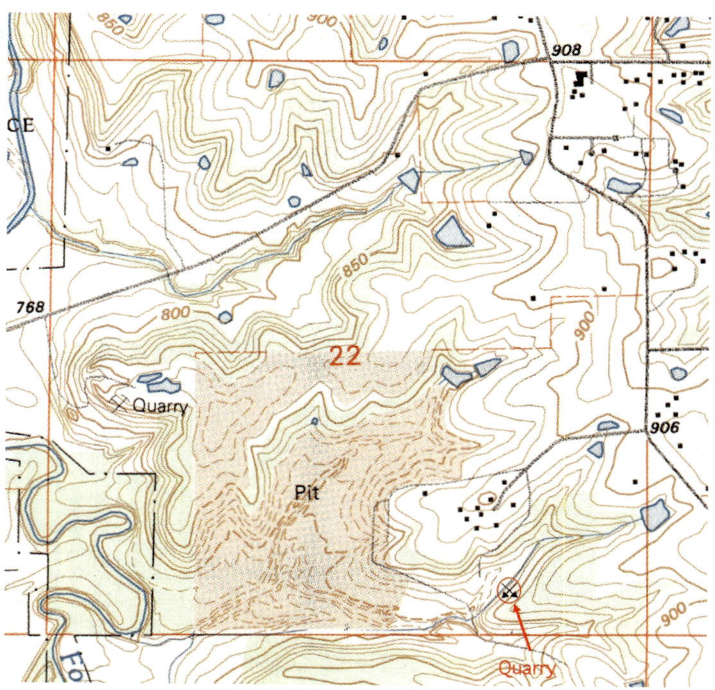

7. Using maps provided by your instructor, locate as many spots as possible in the time remaining . . . using the Township and Range (PLSS) System. Determine all the locations to the nearest quarter-quarter-quarter.

6

Topographic Maps

In Chapter 5, we examined mapmaking in a historical sense, with emphasis on the logic behind mapmaking, and with discussions of the mechanical needs of the mapmaker. In this chapter, we will address several additional facets of mapmaking that drive most of the world's needs for high-quality, information-packed mapping.

Planometric maps provide a great deal of general navigational information, but the need for more map surface detail led to the development of **contour** or **topographic maps,** which place data on a surface that records the three-dimensional shape of the land's surface. In addition, the need for symbols to aid in the interpretation of map features led to the development of worldwide standardization. Figure 6.1 provides a selection of the most commonly used symbols on maps produced in the United States and Canada.

The shift from two to three dimensions involves the use of **contour lines,** lines on a map that connect all points of equal elevation above sea level. A series of successive contour lines thus form invisible boundaries separating all areas of lower elevation from higher elevation. The contour map in the lower portion of Figure 6.2 provides a vivid comparison with the visual perspective most people use when envisioning a landscape.

The normal visual perspective shown in the top portion of the figure is an oblique perspective view of the type an artist might use to render the view. We see higher and lower areas of the scene, see where slopes are gentle and where they are steep, find streams in the valley, and generally get a good understanding of this bit of countryside.

Ordinary planometric maps record none of those same types of information, but all of the important features of the scene are present in contour maps. Contour lines are lines connecting all points in the landscape of equal elevation; for example, all points at 50 feet or 100 feet will fall on a line indicating that elevation. The elevation being measured may either be above or below sea level.

Contour lines always connect elevations at even intervals (whether in feet or meters), and a given contour line will separate elevations higher and lower. If a given contour line was 500 feet above sea level and the line immediately downslope was labeled 480 feet, then we would see that there was a 20-foot difference between the lines. This difference is termed the **contour interval.**

The **contour interval** on the map you use will be stated in the legend, possibly 5, 10, 20, 50 feet or meters or other intervals chosen to record the maximum amount of useful information for the mapped area. Contour lines are always stated as even multiples; in other words, a map with a 20-foot contour interval would have contour lines labeled in multiples of 20 feet: 20, 40, 60, 80, 100, etc. (Remember that 25 is not a multiple of 20.)

KEY TERMS

- *Benchmark*
- *Contour Interval*
- *Contour Line*
- *Contour Map*
- *Hachure*
- *Index Contour Line*
- *Interpolation*
- *Relief*
- *Topographic (Contour) Map*
- *Topographic Profile*
- *Vertical Exaggeration*

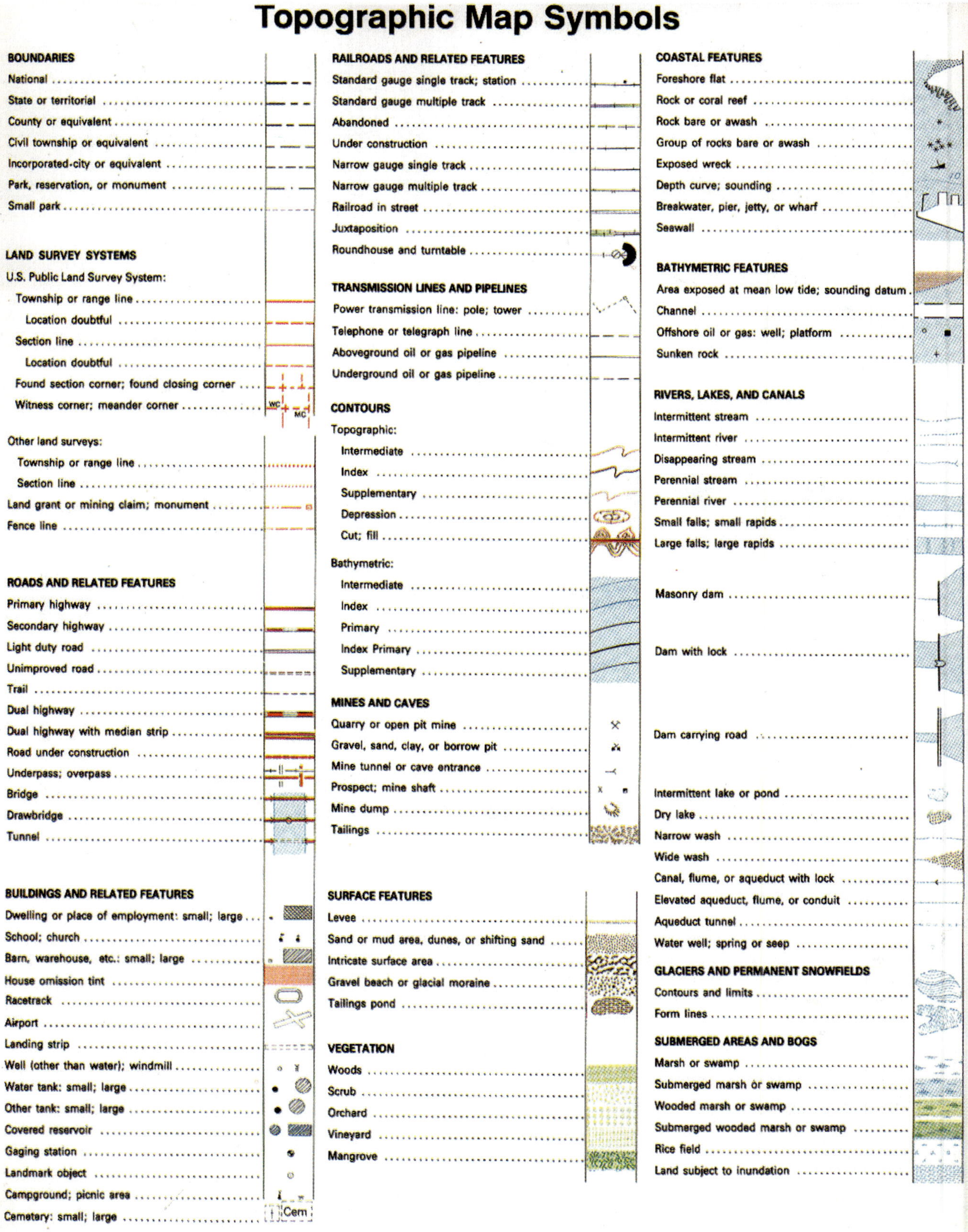

Topographic Map Symbols

BOUNDARIES

National

State or territorial

County or equivalent

Civil township or equivalent

Incorporated-city or equivalent

Park, reservation, or monument

Small park

LAND SURVEY SYSTEMS

U.S. Public Land Survey System:

Township or range line

Location doubtful

Section line

Location doubtful

Found section corner; found closing corner

Witness corner; meander corner

Other land surveys:

Township or range line

Section line

Land grant or mining claim; monument

Fence line

ROADS AND RELATED FEATURES

Primary highway

Secondary highway

Light duty road

Unimproved road

Trail

Dual highway

Dual highway with median strip

Road under construction

Underpass; overpass

Bridge

Drawbridge

Tunnel

BUILDINGS AND RELATED FEATURES

Dwelling or place of employment: small; large

School; church

Barn, warehouse, etc.: small; large

House omission tint

Racetrack

Airport

Landing strip

Well (other than water); windmill

Water tank: small; large

Other tank: small; large

Covered reservoir

Gaging station

Landmark object

Campground; picnic area

Cemetery: small; large

RAILROADS AND RELATED FEATURES

Standard gauge single track; station

Standard gauge multiple track

Abandoned

Under construction

Narrow gauge single track

Narrow gauge multiple track

Railroad in street

Juxtaposition

Roundhouse and turntable

TRANSMISSION LINES AND PIPELINES

Power transmission line: pole; tower

Telephone or telegraph line

Aboveground oil or gas pipeline

Underground oil or gas pipeline

CONTOURS

Topographic:

Intermediate

Index

Supplementary

Depression

Cut; fill

Bathymetric:

Intermediate

Index

Primary

Index Primary

Supplementary

MINES AND CAVES

Quarry or open pit mine

Gravel, sand, clay, or borrow pit

Mine tunnel or cave entrance

Prospect; mine shaft

Mine dump

Tailings

SURFACE FEATURES

Levee

Sand or mud area, dunes, or shifting sand

Intricate surface area

Gravel beach or glacial moraine

Tailings pond

VEGETATION

Woods

Scrub

Orchard

Vineyard

Mangrove

COASTAL FEATURES

Foreshore flat

Rock or coral reef

Rock bare or awash

Group of rocks bare or awash

Exposed wreck

Depth curve; sounding

Breakwater, pier, jetty, or wharf

Seawall

BATHYMETRIC FEATURES

Area exposed at mean low tide; sounding datum

Channel

Offshore oil or gas: well; platform

Sunken rock

RIVERS, LAKES, AND CANALS

Intermittent stream

Intermittent river

Disappearing stream

Perennial stream

Perennial river

Small falls; small rapids

Large falls; large rapids

Masonry dam

Dam with lock

Dam carrying road

Intermittent lake or pond

Dry lake

Narrow wash

Wide wash

Canal, flume, or aqueduct with lock

Elevated aqueduct, flume, or conduit

Aqueduct tunnel

Water well; spring or seep

GLACIERS AND PERMANENT SNOWFIELDS

Contours and limits

Form lines

SUBMERGED AREAS AND BOGS

Marsh or swamp

Submerged marsh or swamp

Wooded marsh or swamp

Submerged wooded marsh or swamp

Rice field

Land subject to inundation

FIGURE 6.1 Standard U.S. Geological Survey map symbols.

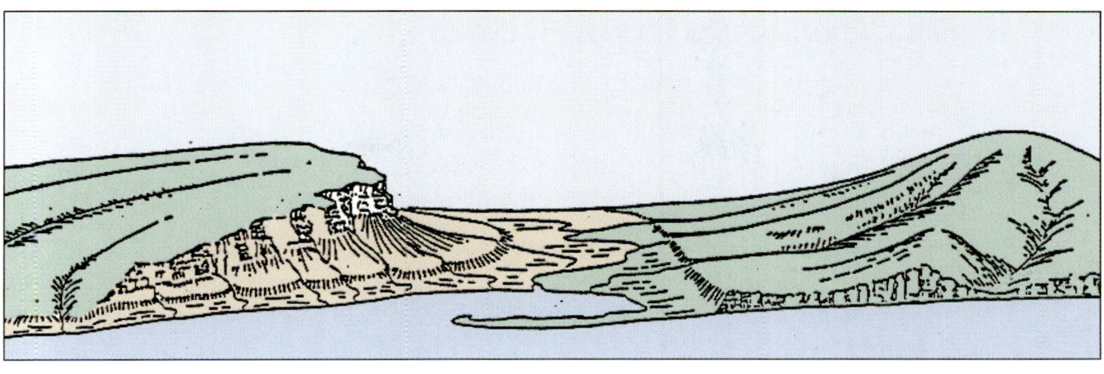

FIGURE 6.2 Comparison between visual perspective of a landscape and a topographic map of the same area.

The contour interval chosen for your map will depend on how flat or hilly the region is (i.e., how much **map relief** is present, relief being the difference in altitude between the higher points on the map and the lower points). Mapping Mount Everest in five foot contours would be silly, as would mapping the nearly flat surface of western Kansas in one hundred foot contours. In each case the map would be nearly useless.

There are two general contour line styles on the map. One is printed in a darker and wider line and is the ***index contour.*** An index contour normally occurs every fifth elevation interval and is labeled frequently. Use the index contour to find out the general elevation of the map area of interest.

In areas of low relief, contour intervals might be small, perhaps 5 or 10 feet between lines, while in regions of tall mountains the contour interval might need to be 20 or 50 feet. The intent is to provide the smallest contour interval that will yield the maximum information.

Between index contours are lighter and thinner lines, which are the **ordinary contour lines.** These are seldom numbered, but by noting the elevation of the index contour above and below the line in question, you can determine the elevation of an unknown point.

■ RULES FOR USING CONTOUR LINES

1. Every location on a contour line connects all points of equal elevation.

2. Contour lines separate points of higher elevation (uphill) from points of lower elevation (downhill). To distinguish between uphill and downhill, look at the adjacent contour lines and check their elevations.

3. The difference in elevation between any two adjacent contour lines is the contour interval. Index contours are printed as heavier and darker lines, usually every fifth contour line, and are labeled with elevation numbers in feet or meters.

4. Contour lines always close or come back to the starting point so that they form continuous loops (a contour line cannot just end) and form irregular loops. Note that loops will often extend off the map, and you might not be able to trace the entire closure on one map.

5. Contour lines never really cross since each connects only points at a given elevation. On a map they can appear to cross only in the case of an overhanging cliff when the lower slopes are recessed. In this very rare case, the lower hidden contour lines are dashed.

6. Contour lines can merge only in the case of a vertical slope such as a cliff. They do not in fact merge, but with a vertical cliff the successive contour lines would be drawn so that one was directly on top of the other and they would simply appear to have merged.

7. Evenly spaced contour lines suggest a uniform slope.

8. Closely spaced contour lines suggest steep slopes while widely spaced contours record gentle slopes.

9. A series of concentric, closed contour loops will define a hill or mountain.

10. Concentric closed contour loops with **hachures** (short dashed lines at right angles to one side of the contour line) indicate a closed depression. The hachure marks point toward the deeper part of the depression.

11. If streams flow down the area being mapped, contour lines will form a "v" pattern where they cross the stream, and the "v" will point upstream (uphill).

12. If mapped contour lines forming a "v" pattern point down hill, the contour lines are defining a ridge.

Before a contour map is prepared, the area must be surveyed to find the highest and lowest elevations present. The difference between the high and low areas of the map is the **relief.** **Local relief** is the difference in elevation between two areas being compared while **total relief** is the difference in elevation between the highest and lowest areas on the map.

Contour intervals also allow one to determine the **gradient,** the expression of the steepness of a slope. **Gradient** is determined by dividing the relief between two points of interest by the distance measured from the map and is expressed in feet per mile or in meters per kilometer.

■ TOPOGRAPHIC PROFILES AND VERTICAL EXAGGERATION

Yet another way to discover the shape of landforms in an area is to prepare topographic profiles. *Topographic profiles* are cross-sectional representations of the high and low areas of a map. To draw a topographic profile from a contour map, first select a suitable cross section such as A–A' (Figure 6.3). After the cross section has been chosen, points where contour lines cross the A–A' line are transferred to a graph with a vertical scale.

In transferring the data points, fold a piece of blank paper and line it up with the cross-section line on the contour map. Make marks on the paper where each contour crosses the A–A' line and at each end of the section line. Record the elevations at your contour marks.

Transfer marks from the paper to the vertical profile. The vertical scale includes maximum and minimum contour lines measured from the map. Match up elevations on your paper with elevations on the cross section by moving the paper up and down the graph until the elevation point matches the elevation line and mark that point on the graph.

Once all points are marked, connect the dots with a smooth line (remember—on a real landscape, almost all surfaces are curved). Be on the "lookout" for typical topographic features, such as hills, valleys, depressions, etc. Draw hills and valleys to resemble natural landscapes (Figure 6.3).

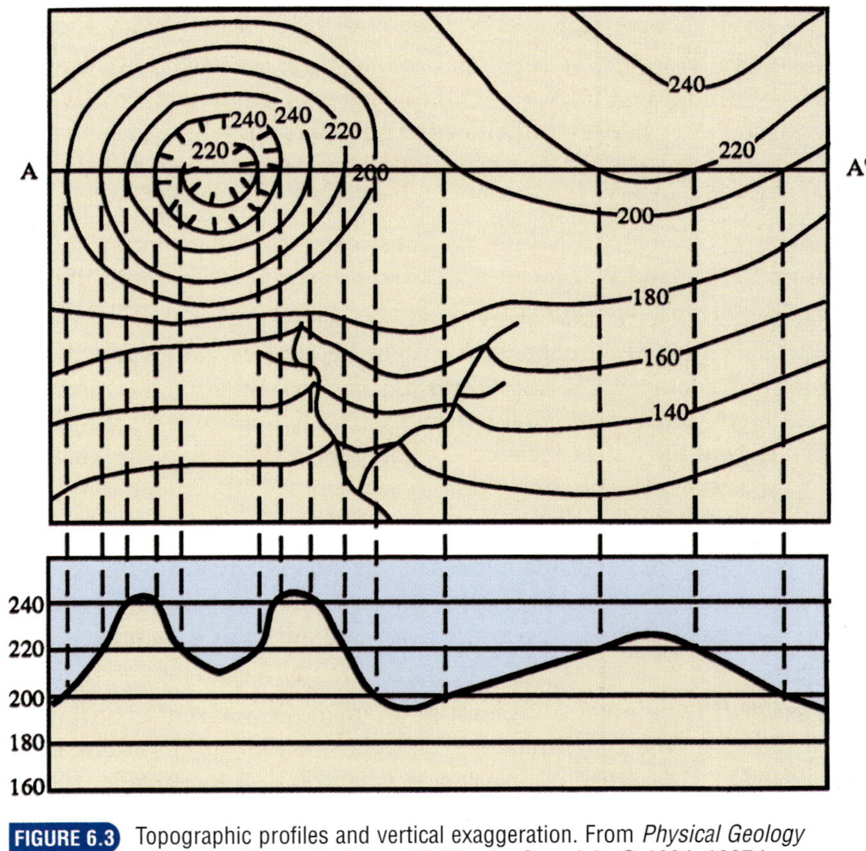

FIGURE 6.3 Topographic profiles and vertical exaggeration. From *Physical Geology Laboratory Manual* by Karen M. Woods. Copyright © 1994, 1997 by Kendall Hunt Publishing Company. Reprinted by permission.

Vertical exaggeration of a ***topographic profile*** is the amount the profile has been stretched vertically. The vertical scale is exaggerated to emphasize the shape of hills and valleys since in almost all areas the natural height of the land changes so slightly that a true vertical profile would show little evidence of hills and valleys.

Vertical exaggeration is represented as a ratio of one inch measured along the horizontal map scale in feet, divided by the number of actual feet in one measured inch on the vertical scale. For example, if the horizontal (map) scale states that one inch on the ground = 500 feet, and one measured inch on the vertical scale equals 250 feet, divide 500 feet by 250 feet. The vertical exaggeration will be 23 (two times). Vertical exaggeration taken from a map must always be greater than one.

■ MAKING YOUR OWN CONTOUR MAP

Data sets needed to prepare contour maps can be gathered in several ways, but, historically, the most common way has been to send out a survey team consisting of a surveyor and an assistant (generally referred to as a "rodman"). These survey team members locate a series of spots on their base map and indicate the elevations above sea level at those sites. Once the survey is complete and the information is brought back to the office, a mapmaker then draws in the actual contour lines. Today, much mapping is done with overlapping aerial photos or even with photographic data from orbiting spacecraft.

The numbered elevations on your map are not necessarily the highest or lowest spots. Labeled elevations on maps sometimes consist of just a number and sometimes a number with an "x" beside it **(spot elevations).** Other times there may be a small triangle with a dot labeled ***BM*** for ***benchmark.*** These are surveyed elevations of varying degrees of accuracy.

Benchmarks (sites labeled BM) are the most carefully determined elevations and are marked on the ground by bronze plates fixed to concrete bases at the exact site where the elevation was determined.

Sometimes surveyed elevations may have just been placed at a spot convenient to the survey team. Check contour lines to make sure that you are not fooled into missing the actual highest or lowest point by simply reading numbers off the map. Check contour lines to find the actual locations of high and low spots. Who knows, maybe the survey team was lazy—and only went to easy spots and avoided climbing to the tops of the highest hills.

Often the point you are interested in does not lie directly on one of the contour lines, and its elevation must be estimated. In the exercises at the end of the chapter, you will have an opportunity to estimate the locations of contour lines by a process known as interpolation.

In ***interpolation,*** the precise location of an unknown point of elevation is determined by estimating where it would fall between the known elevations higher and lower. For instance, if you had one surveyed data point on the map at 160 feet and another at 200 feet, while the map contour interval was stated to be 20 feet, it is evident that the 180 foot contour location is located halfway between the 160 and 200 foot marks. Likewise, perhaps the survey team located elevations at 32 feet and 66 feet, and your map uses contour intervals of 20 feet (Figure 6.4). The relief (vertical difference between the two points) is 34 feet. Therefore, by interpolation, we can see that the 40 foot point will be approximately one-quarter of the distance from the lower to the

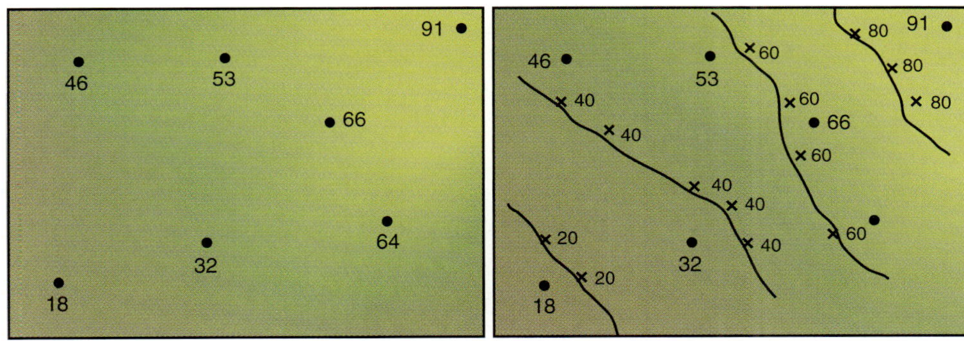

FIGURE 6.4 Determination of map elevations by interpolation to find the 20-foot contour intervals.

higher point, and the 60-foot point will be proportionately below the 66-foot data point. The locations determined by this interpolation method will provide good places to locate your contour lines.

The sketches in Figures 6.5, 6.6, 6.7, and 6.8, illustrate ways to handle common features of landscapes when preparing your contour map. (From *Physical Geology Laboratory Manual* by Karen M. Woods. Copyright © 1994, 1997 by Kendall Hunt Publishing Company. Reprinted by permission.)

The margins of quadrangle maps record many types of data around the mapped area. Figure 6.9 points out some of the most useful features of quadrangle maps.

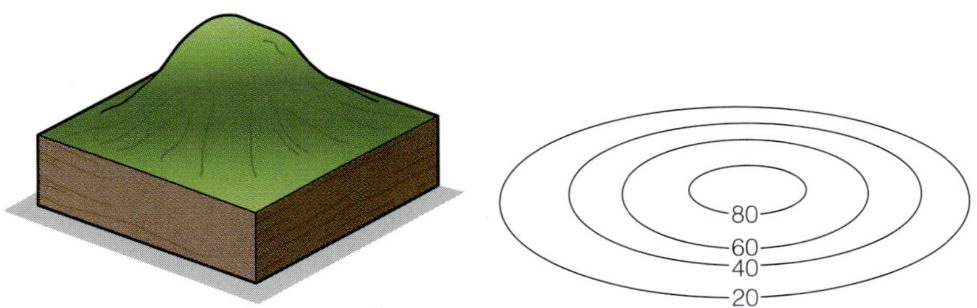

FIGURE 6.5 **Closed contour lines,** indicate hills. The successively smaller rings formed by the contour lines indicate higher elevations.

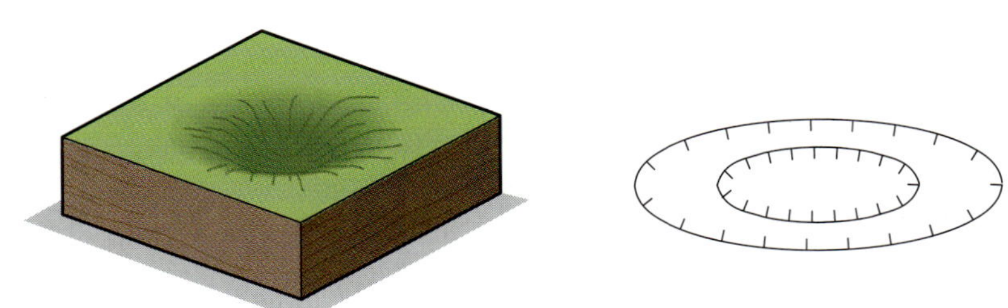

FIGURE 6.6 **Closed contour lines with hatchurs** indicate depressions. The hatchurs point the inward and downward in the direction of the depression.

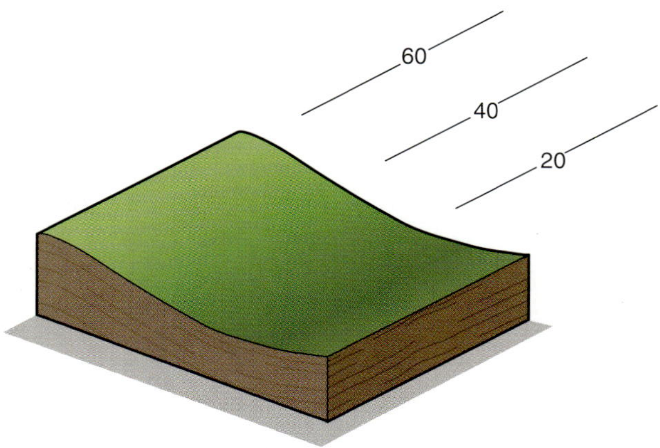

FIGURE 6.7 **More or less parallel and evenly spaced contour lines** indicate slopes. The closer the contour lines, the steeper the slope and vice versa.

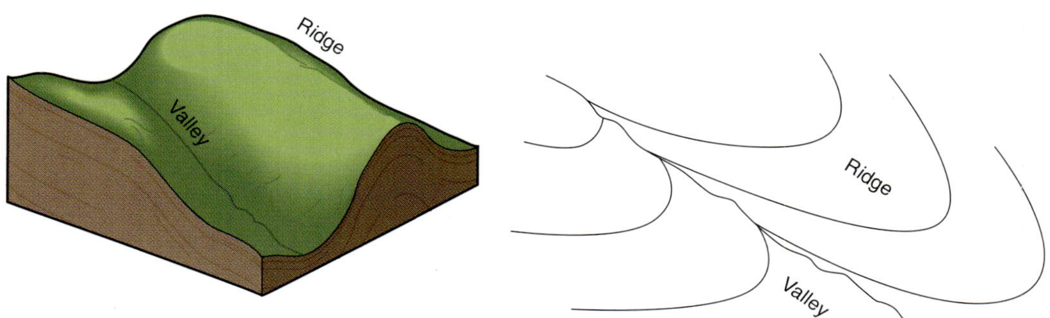

FIGURE 6.8 **Contour loops.** Contour lines that bend upslope or downslope indicate either valleys or ridges. Contour "fingers" that point downslope define ridges, while contour "fingers" that point upslope indicate stream valleys.

◼ THE EDGES OF THE MAP

The edges of a modern contour map, as prepared and distributed by the United States Geological Survey, contain a large array of very useful information. It will make the map chapters of this book much more meaningful if you can glance at the map border and immediately understand what all those numbers and symbols mean. Your instructor can provide some useful maps for this exercise. The examples below will allow you to extract a great deal of very useful data from the edges of your maps.

The margins of most maps contain the following types of information.

Longitude and Latitude

Examine the upper right hand corner of the map in Figure 6.21. Note the number 94° 15' 00". This is the longitude of the east edge of the map. You can confirm this by looking at the lower right hand corner of the map . . . where the same number appears. Therefore since the number does not change, it must be defining the east edge of the map.

There are two numbers in this upper right hand corner of the map. The second number is 39° 07' 30", which is the latitude of the north boundary of the map. You can confirm this by looking at the upper left hand corner of the map (Figure 6.23), where the same number appears, therefore defining the North edge of the map.

Check the same pairs of numbers on all four corners of another map (provided by your instructor), and determine which numbers represent longitude and which represent latitude.

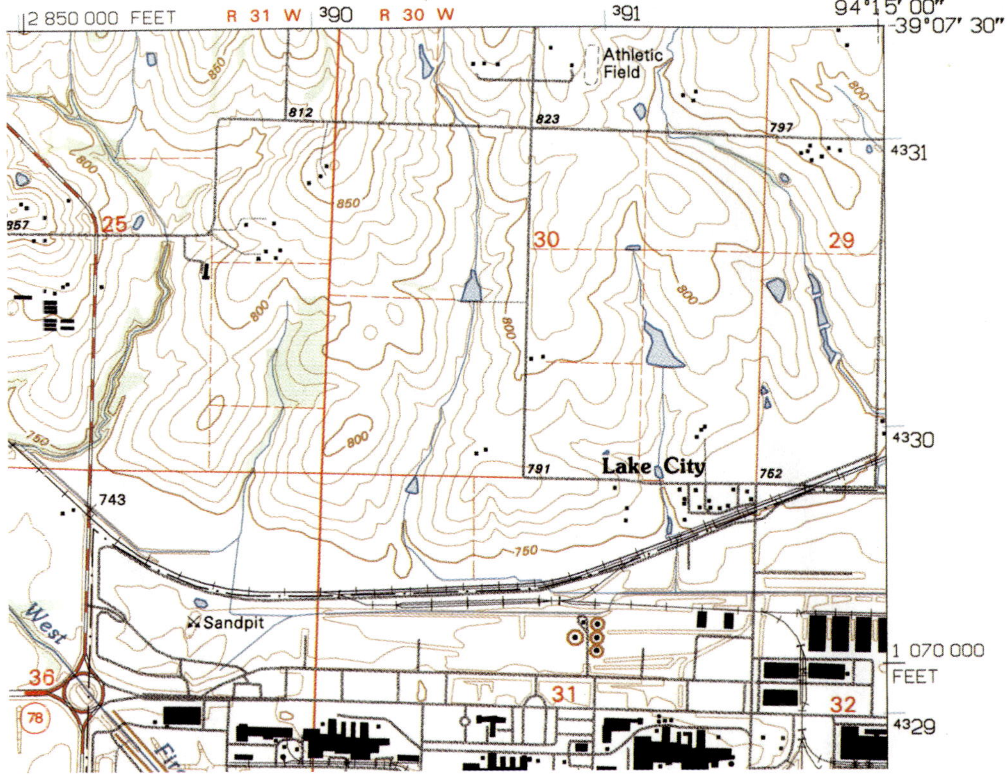

FIGURE 6.21 Longitude and Latitude, upper right hand corner of the map.

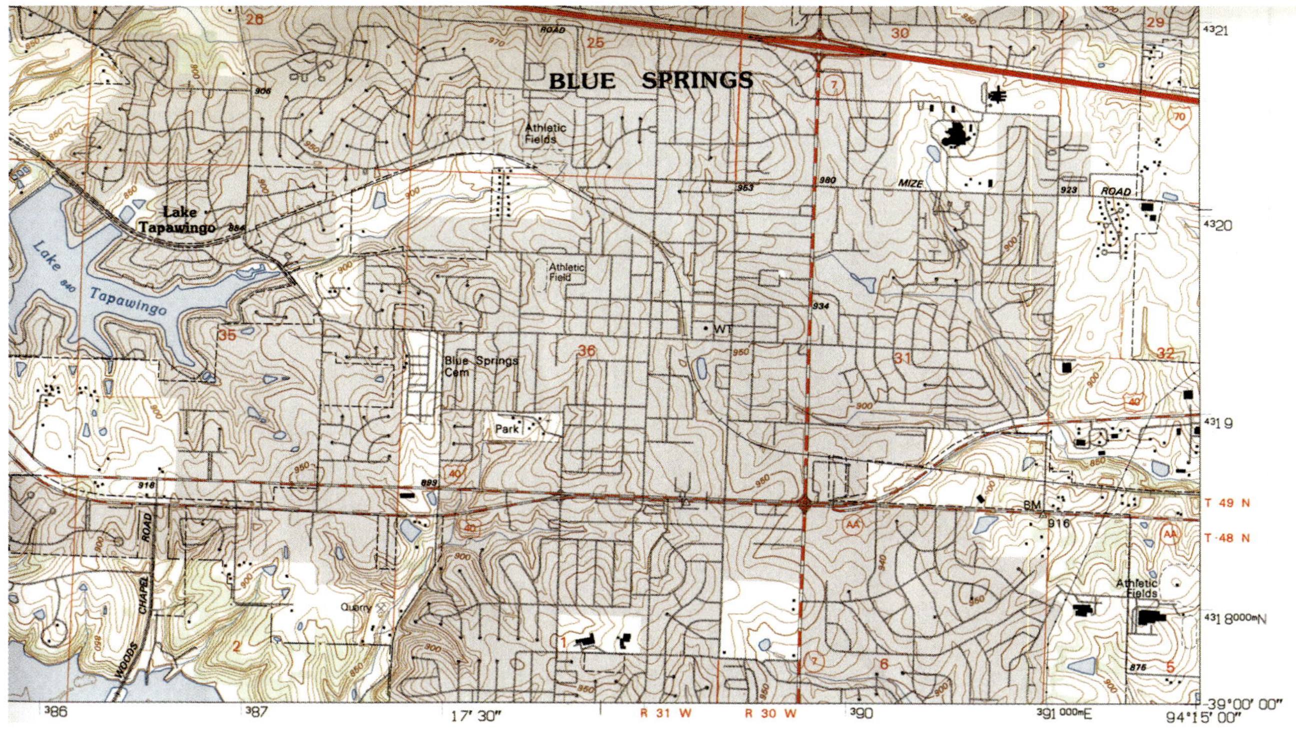

FIGURE 6.22 Longitude and Latitude, lower right hand corner of the map.

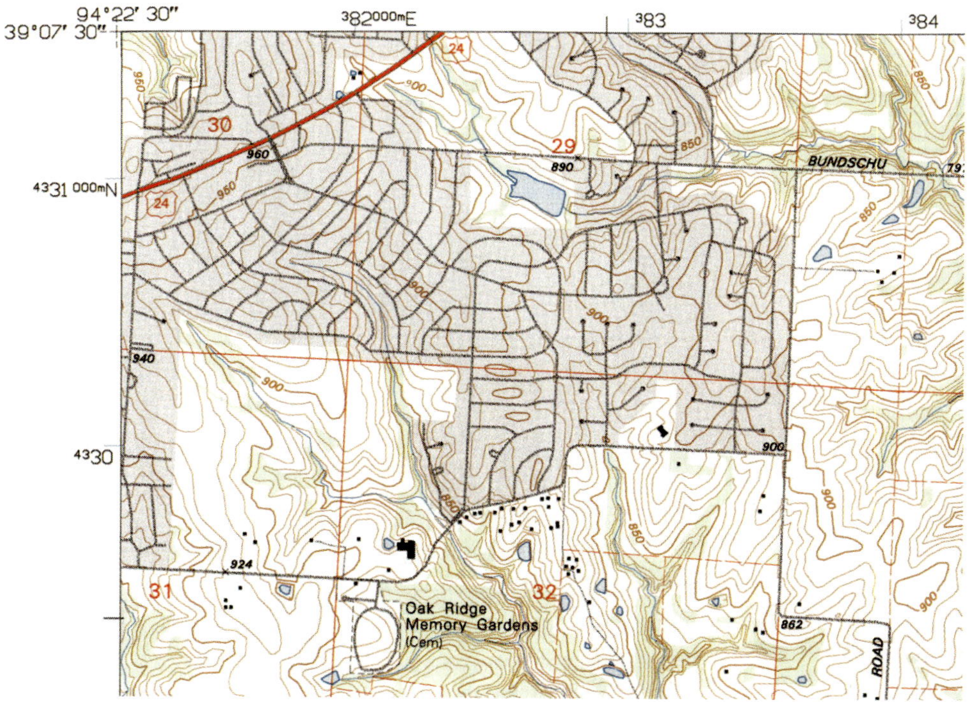

FIGURE 6.23 Longitude and Latitude, upper left hand corner of the map.

There is an additional set of useful numbers for working with longitude and latitude. Go back to the upper left hand corner of the map and look to the right. The numbers 17' 30" appears one third of the way across the map margin (Figure 6.24), and another third of the way to the right you will see 20' 00". These index marks along the top and bottom edges, and the right and left edges of the map help you locate the longitudes and latitudes with greater precision, dividing each edge of the map into thirds. All four sides of the map have similar markings.

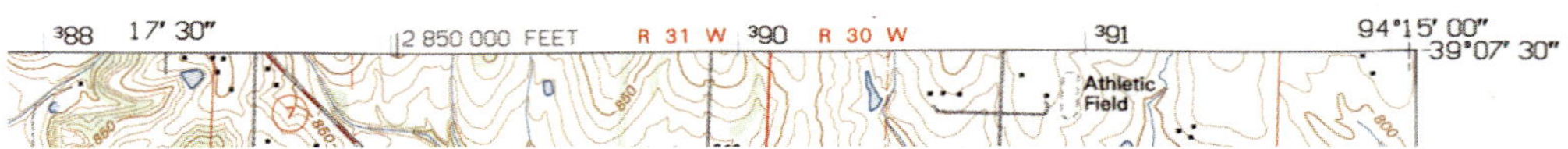

FIGURE 6.24 Longitude and Latitude, index marks dividing the edge of the map in thirds.

Township and Range (PLSS) Data

Red letter and number sets appear at intervals along the margins of the map, such as T. 49 N. and T. 48 N. or R. 30 W. and R. 31 W. (Figure 6.25)

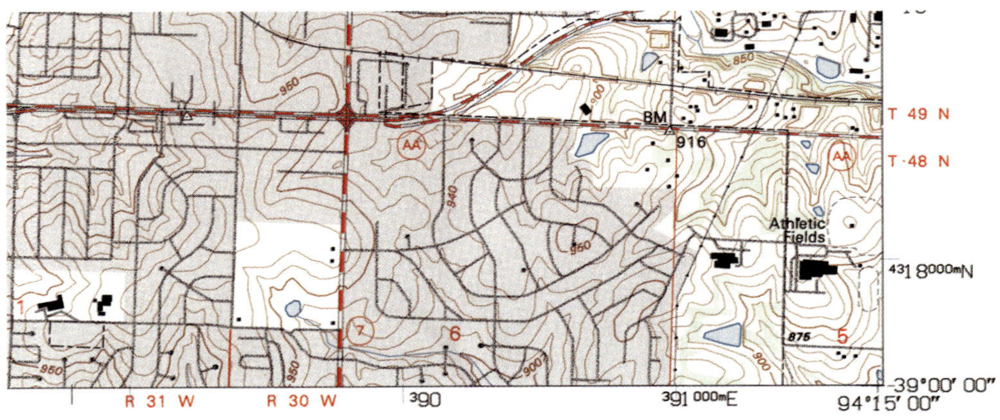

FIGURE 6.25 Township (or Tier) and Range name markers.

Map Scales

There are at least three types of scale information on this map.

Longitude and Latitude Upper right hand corner of the map: '7.5 Minute Series' This information states that the map is 7' and 30" (seven minutes and thirty seconds) of longitude from East to West, and 7' and 30" of latitude from North to South. Remember, half a minute equals 30" (seconds). Other maps might be in different scales, such as 15' (fifteen minute) or larger.

Ratio Scale (Fractional Scale) In Figure 6.26, an image located in the center of the lower margin of the map, the number '1:62,500' appears. As stated on page 103, this number is the ratio of some measurement on the map compared to 62,500 of them on the ground. For instance one inch measured on the map = 62,500 inches on the ground. This can be measured in any units, inches, centimeters or any other arbitrary units.

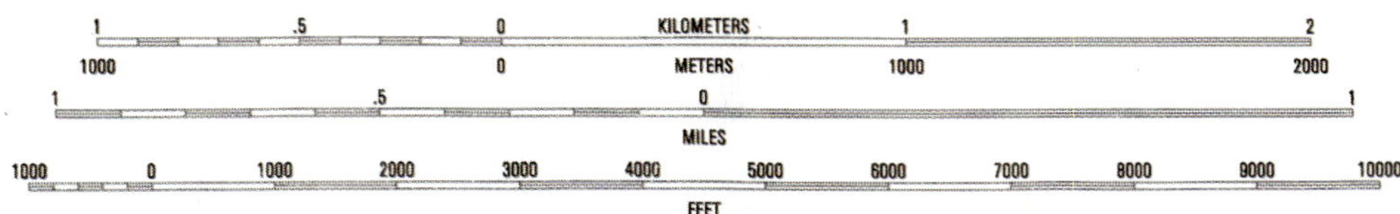

CONTOUR INTERVAL 10 FEET
SUPPLEMENTARY CONTOUR INTERVAL 5 FEET
NATIONAL GEODETIC VERTICAL DATUM OF 1929

FIGURE 6.26 Ratio and Bar Scales, plus the contour interval.

Bar Scale (Graphical Scale) Horizontal bars below the Ratio Scale are commonly subdivided into three sets of scales that can be used like rulers to measure distances on the map surface. The units commonly used are miles, feet and kilometers (Figure 6.26).

Contour Interval The contour interval will be stated in the lower central edge of the map. Units of measurement may vary, as the map may be an older one using feet, or a newer version using metric measurements (Figure 6.26).

Magnetic Declination In the lower edge of the map, generally to the left of the bar scales, is a symbol that allows you to see how the map's North orientation differs from the orientation read from your compass or from a UTM map grid.

The central line usually has a star at the upper end of the line. This may or may not be otherwise labeled, but it represents the direction of true, or **geographical North**. Geographical North is the point on the globe where all the lines of longitude converge (Figure 6.27).

The second line in this symbol may be either on the left or right side of the North indicator with the star. This will vary with your location, but the line will normally end in an arrowhead symbol and be labeled MN, for **Magnetic North**. Alongside this MN line will be a number, such as 5½° (five and a half degrees). This number is very important, because it tells you that the North arrow of your compass is pointing 5.5° away from true north. You will need to add or subtract 5.5° from your compass reading to get the correct North direction.

There is a third line, normally terminating in 'GN'. This is the deviation from true North of the UTM map grid, and is called **'Grid North'**. This will be further explained in Chapter 8.

Do these arrows really matter? Yes, indeed. Consider the different orientations in the following examples (Figure 6.28).

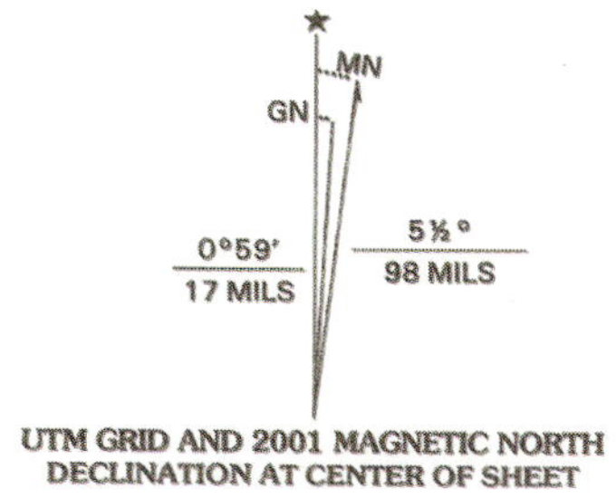

FIGURE 6.27 Magnetic Declination for Hurst, Texas, a suburb of Dallas, TX.

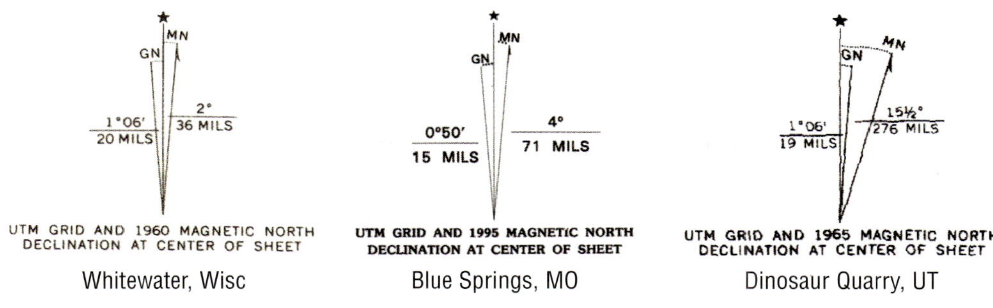

FIGURE 6.28 Representative magnetic declinations.

With the shifting of lithospheric plates and the slow change in location of the magnetic poles, the magnetic declination is also being constantly changed. Older maps will have incorrect data, and must be corrected for today's orientations.

How bad can it be if you just ignore the magnetic declination and assume your compass is pointing to true geographical north? Consider this general formula: one degree of compass error equals nearly 100 feet of ground error.

Let's assume that your map showed a magnetic declination of 13°. Those 13 degrees of difference between true north and magnetic north, and this difference will result in 1,300 feet (more than 400 yards) of error for every mile traveled. You'll be just under a mile off for every four miles covered (i.e. assuming you followed your compass without adjusting for magnetic declination). In this example, you would end up far away from where you would have been if you walked toward true north. In the right scenario, this could be disastrous, perhaps even fatal!

UTM Mapping Data

Odd number combinations printed in upper and lower case combinations are also seen, such as $^{43}18^{000m}$N. These measurements are used in UTM (Universal Transverse Mercator) mapping and will be explained further in Chapter 8. Basically, they are measurements in meters from index points (Figure 6.29).

The most important point here is that you do not confuse the UTM markings with PLSS, or Longitude and Latitude markings.

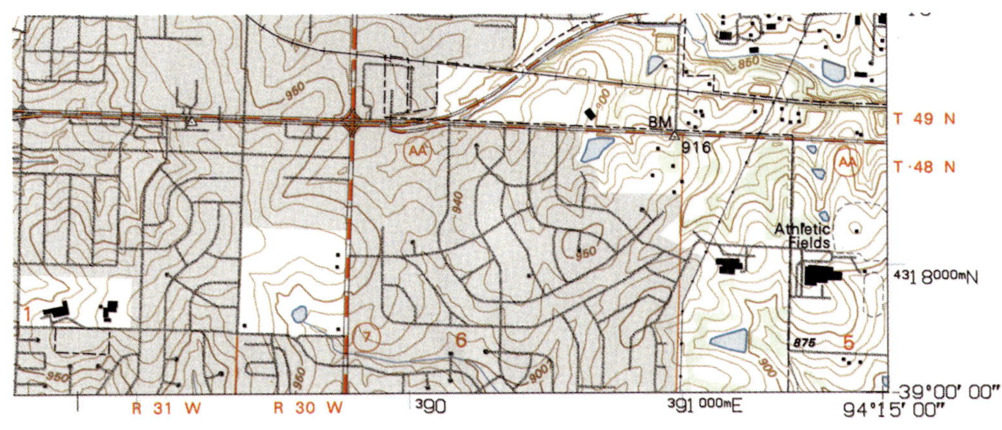

FIGURE 6.29 UTM markings.

State Plane Coordinate System   A map projection that measures distance in feet. By providing an east measurement and north measurement, the state name, and the zone number, any location in the United States can be identified by a unique coordinate value.

The State Plane Coordinate System is a set of 126 geographic zones. Each state contains one or more state plane zones. There are 110 zones in the continental US, with 10 more in Alaska, 5 in Hawaii, and one for Puerto Rico and US Virgin Islands (Figures 6.30 and 6.31).

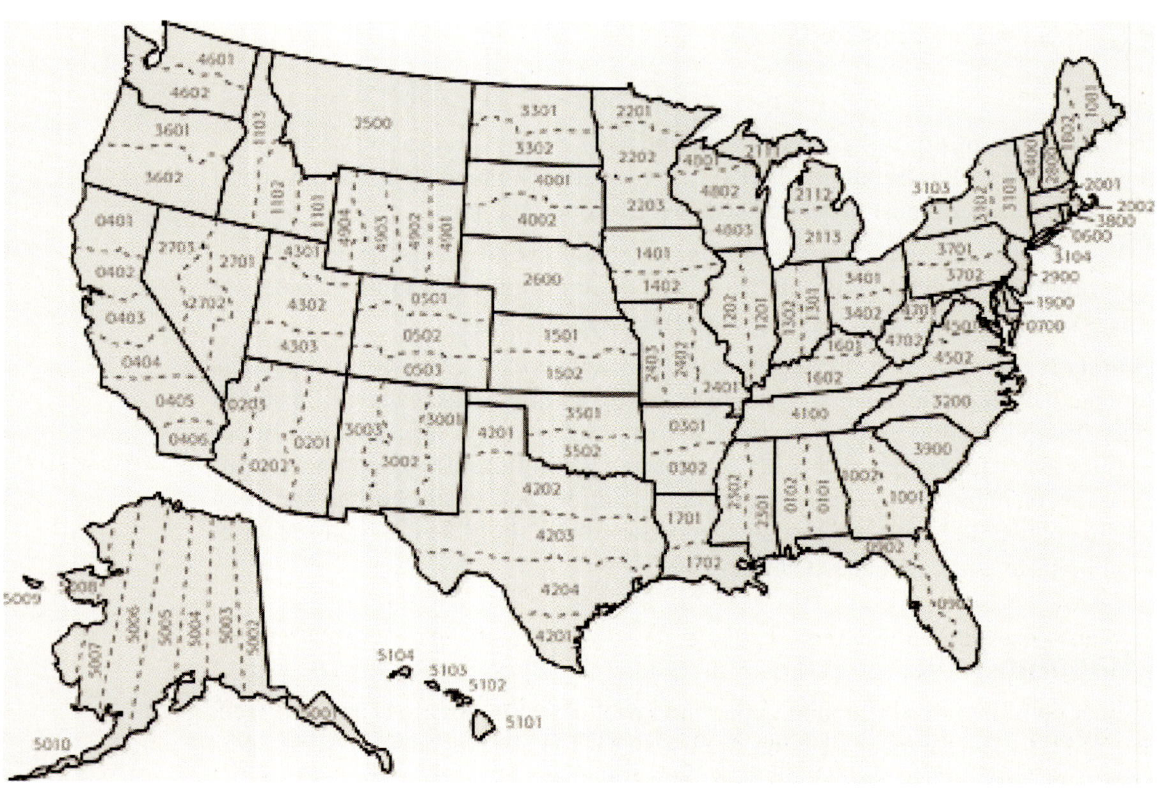

FIGURE 6.30   State Plane Coordinate System zones for the United States. From www.pobonline.com. Used with permission.

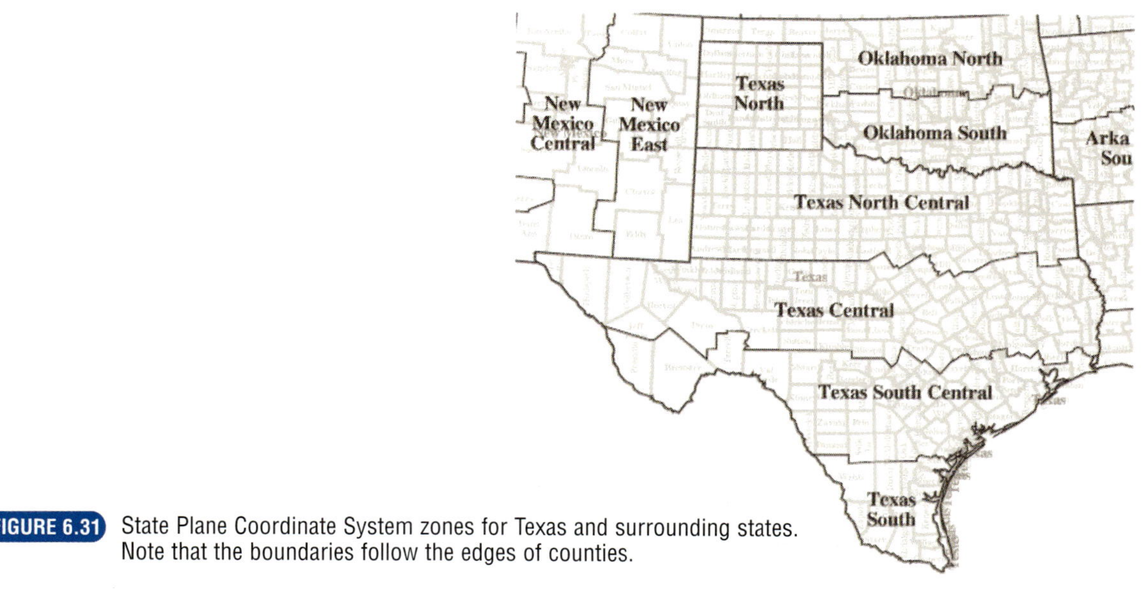

FIGURE 6.31   State Plane Coordinate System zones for Texas and surrounding states. Note that the boundaries follow the edges of counties.

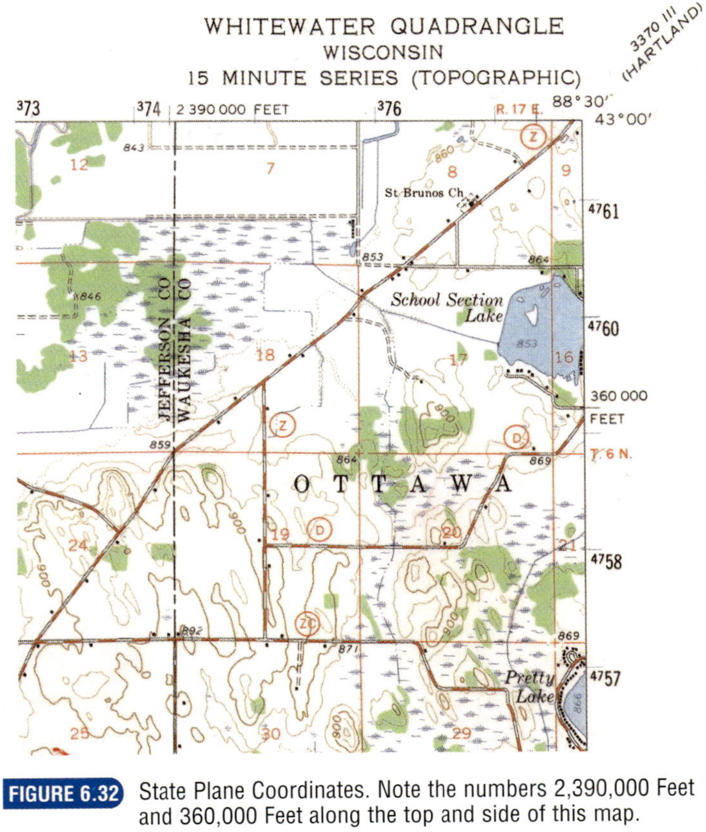

FIGURE 6.32 State Plane Coordinates. Note the numbers 2,390,000 Feet and 360,000 Feet along the top and side of this map.

The system uses a simple Cartesian coordinate system to specify locations rather than a more complex system of latitude and longitude. By ignoring the curvature of the Earth, "plane surveying" methods can be used, speeding up and simplifying calculations. Second, the system is highly accurate within each zone (error less than 1:10,000). The main problem with the system is that each zone uses a different coordinate system.

Locations are specified along the margins of the map in feet or meters from precisely define index points (Figure 6.32). This system is widely used for common survey work. In the example in Figure 6.32 (above), the number on the top of the map, 2,390,000 Feet, and the number on the right map margin, 360,000 Feet, are State Plane Coordinate numbers.

General Map Names and Orientation

Other markings can be very useful. The official map quadrangle name, required for identification and map ordering, will be found in the upper right hand corner of the map (see Figure 6.33). Another statement in the upper left hand corner of the map identifies the agency that prepared the map (Figure 6.34).

BLUE SPRINGS QUADRANGLE
MISSOURI-JACKSON CO.
7.5 MINUTE SERIES (TOPOGRAPHIC)

FIGURE 6.33 Upper right hand corner: Formal name of the map.

UNITED STATES
DEPARTMENT OF THE INTERIOR
GEOLOGICAL SURVEY

FIGURE 6.34 Upper left hand corner: Name of the agency that prepared the map, i.e., 'United States Department of the Interior, Geological Survey'.

In the lower left hand corner of most maps is a block of data that records for map's history (Figure 6.35)

Should you wish to continue your survey beyond the margins of the map, you will need to know the names of the adjacent maps sheets in order to order them from the appropriate agency. Figure 6.36 shows an example of the most modern index method. Older maps will have the adjacent map names located in the centers of each map margin, and printed diagonally from each of the four corners.

Now that you can read the edges of the map, practice these new skills on some complete maps provided by your instructor.

There is a great Internet site generated by the National Geophysical Data Center, a division of NOAA (National Oceanic and Atmospheric Administration). This site will be found at **http://www.ngdc.noaa.gov/geomagmodels/struts/calcDeclination**.

Type in the zip code of your home and it instantly provides the longitude and latitude. If you click on the 'Compute Declination' button, the site generates a map of your home area and superposes a compass rose on your dwelling. Further, this site computes the magnetic declination and tells you how fast it is changing. For the author's home, this variation was: Declination = 4° 16' E, changing by 0° 7' West per year. See what these values are for your home town, or for some place you have visited.

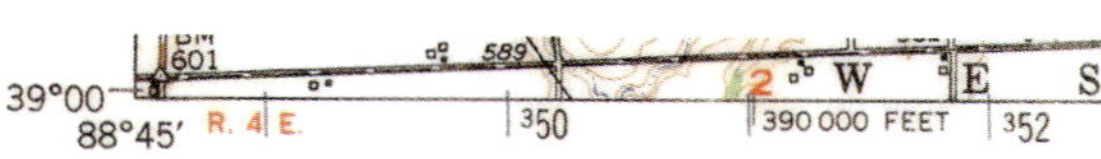

Mapped, edited, and published by the Geological Survey

Control by USGS and USC&GS

Culture and drainage in part compiled from aerial photographs taken 1938–1939
Topography by plane-table methods 1949–1950

Polyconic projection.   1927 North American datum
10,000-foot grid based on Illinois coordinate system east zone

Red tint indicates area in which only landmark buildings are shown

1000-meter Universal Transverse Mercator grid ticks, zone 16, shown in blue

FIGURE 6.35 Data concerning the history, times when corrections were made, and methods of generation of this map.

QUADRANGLE LOCATION

1	2	3
4		5
6	7	8

1 Liberty
2 Missouri City
3 Buckner
4 Independence
5 Oak Grove
6 Lees Summit
7 Lake Jacomo
8 Tarsney Lakes

ADJOINING 7.5' QUADRANGLE NAMES

FIGURE 6.36 Index to adjacent maps. Names of the maps found to the immediate Northwest, Northeast, Southwest and Southeast of the present map. For example, the next map to the Northwest of this sheet is the 'Liberty, Missouri' quadrangle map.

Pace and Compass Mapping
Finding Your Way with Map and Compass

In this chapter, you will learn to use a compass as a useful tool for simple surveying and for locating yourself during hikes and other trips. Compasses can be simple or very sophisticated, so the first step is to learn the nature of a compass.

Because geometry has divided circles into subunits, the compass also employs those same divisions. Therefore, the edge of a compass is divided into 360 degrees. Look at your compass face and you will see that both 0 degrees and 360 degrees are at the top of the compass dial, while 90 degrees is half way down the right side, 180 degrees is at the bottom, and 270 degrees is half way up the left side . . . and when using a map, don't forget that the top of the map is north!

While hiking or mapmaking, your compass will be held so that the top (360/0 degrees) is pointing north. The sections between the other major divisions are termed **quadrants;** therefore, you would describe the area between 0 and 90 degrees as the northeast quadrant, between 90 and 180, the southeast quadrant, and so forth around the dial (Figure 7.1).

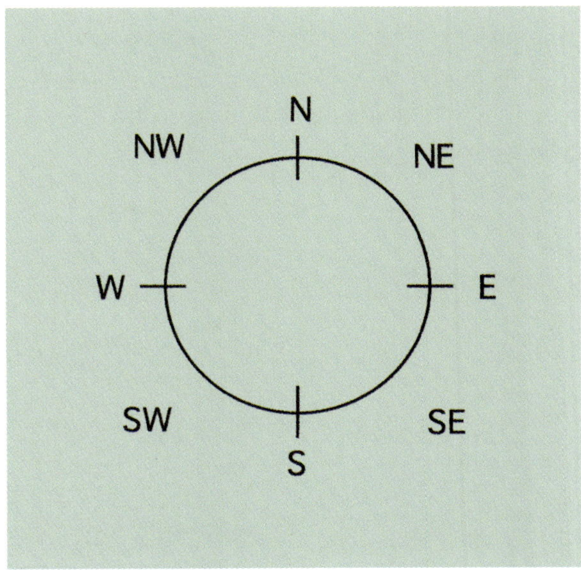

FIGURE 7.1 Map orientation in quadrants.

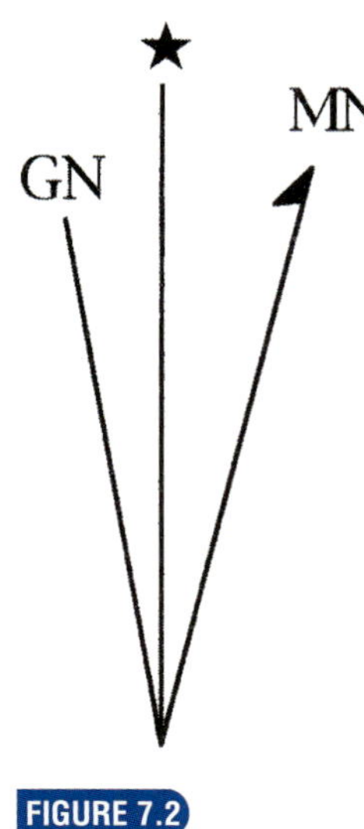

FIGURE 7.2

IMPORTANT: A simple compass consists of a magnetized metal needle that is supported on a pivot point so that it can move easily no matter how the compass is reoriented. The needle is able to align itself to the magnetic field lines of the earth.

Your compass seldom points to true, or **geographic north,** but always points to the location of **magnetic north.** These are NOT the same locations, for magnetic north is currently located northwest of Hudson Bay in northeastern Canada. If you were standing at a location straight south of that site, your compass would correctly point due north, but, at nearly every other location on earth, the compass would not point to geographic north (Figure 7.2).

This difference between true geographic north and magnetic north is termed the **magnetic declination** and must be accounted for when using a compass, or you will not arrive at the desired site.

Topographic maps will show a symbolic indication of true and magnetic north at the bottom edge of the map. The straight arrow, commonly ending with a star, is true geographic north. GN indicates Grid North (see Chapter 6), while the MN line is the direction to Magnetic North. The number indicated is the difference in degrees between true and magnetic north.

What happens when you just ignore these corrections? After all, how much difference can ten or twelve degrees make, anyway? Well, look at Figure 7.3, and you'll see that after a ten-mile trip, by ignoring a ten-degree correction you'd be off over 1.7 miles! If you were lost and out of supplies, such an error could be a lot more than inconvenient; it could be fatal!

◼ COMPASS PARTS

Nearly all compasses have the same fundamental parts: base plate, straight edge and ruler, direction of travel arrow, compass housing with 360 degree markings, North label, index line, orienting arrow, and a magnetic needle (north end is red) (Figure 7.4).

Your compass housing has a moveable outer ring with degree marks, and your instructor will show you how to adjust the compass so it will read correctly and indicate true north for your location.

Declination correction in degrees	Location error after ten miles
1 degree	920 feet (280 meters)
5 degrees	4,600 feet (1,402 meters)
10 degrees	9,170 feet (2,795 meters)

FIGURE 7.3 Failure to include magnetic declination corrections.

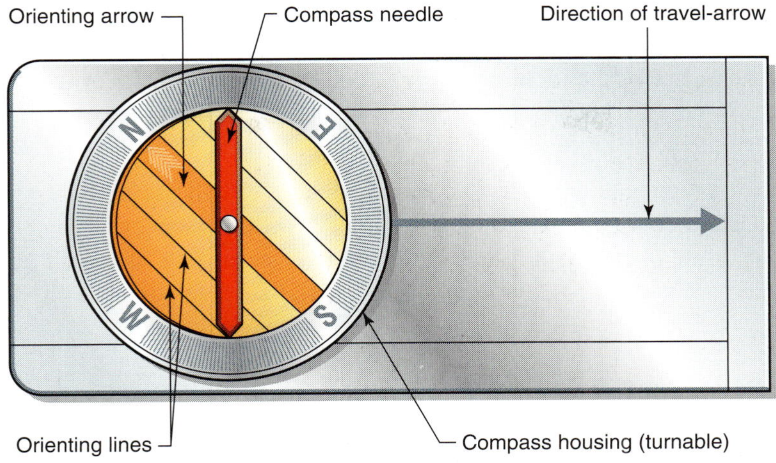

FIGURE 7.4 General features found on compasses.

Once your compass has been adjusted to account for **magnetic declination,** you can use it to determine the **bearing** or **azimuth** between any two points on the earth.

The **bearing** can be simply defined as the direction from one location to another. Bearings are expressed in degrees either east or west of true (geographic) north or south. To determine your bearing, locate two points of interest on your topographic map and connect them by lightly drawing a pencil line between them (remember, lightly—you will need to erase this line later without damaging the map surface). It will be very helpful to extend this line to the edge of the map.

Bearings can now be determined with either a simple protractor or with a compass. Place either instrument along the edge of the map (this will allow you to be sure that you have located true north—remember that the right and left edges of a map are north-south lines). With the 0 degree mark of your compass toward the map top and the 180 degree mark toward the bottom, read off the number of degrees between the 0 degree mark and the intersection with the line you drew on the map. If this number is 23 degrees and the line you drew was toward the left margin of the map, then the bearing would be read as North 23° East.

Map directions can also be read as *azimuth bearings* or simply as *azimuths.* In this case, the degrees read omit the compass directions (north, northeast, southwest, etc.) and simply relate the reading to a 360° circle. Therefore, a measurement to the southwest that might have been said to have a bearing of South 20° West would also have an azimuth of 200°.

Why have azimuths? Reading from a 360° circle is less confusing for some persons who might not remember which way was west or north. To some degree, the system you choose to use will be determined by personal preference.

PACE AND COMPASS MAPPING

If you are an avid hiker or if you need a quick map of your property, pace and compass mapping can be very useful. This system allows accurate location determination for sites of interest. Before starting on your map, examine your compass and make sure that the proper corrections for magnetic declination have been made and checked.

Before the map can be started, you also need to know the length of your average walking pace. With the help of a lab partner, walk at a normal and comfortable pace while the partner marks the location of your feet (this is done easily on a sidewalk with chalk). Determine the length of that stride in inches or centimeters from the point where your right foot first touched down to the spot where the left foot touched.

With a little practice and awareness, this stride length will prove to be very uniform over smooth ground. If you are going up or down hills, the pace length will change and should be accounted for in your measurements. If your campus has a hill nearby, measure the pace on a slope and compare it with the flat surface pace length.

■ EXERCISES

Pace and Compass Mapping Project

In order to accomplish this project, you must learn three new skills:

1. Pacing for distance

2. Taking compass bearings for directions

3. Plotting data sets on graph paper

Part One: Pacing and Determination of Pace Length

To determine your individual **PL (pace length),** the instructor will have measured off a selected distance (a section of sidewalk, a classroom hallway, etc.). Each student will individually start at one end and pace to the other, using a comfortable pace length and counting the steps from one end to the other of the reference distance. Students should pace along this reference distance a second and even a third time to help establish consistency in determining the actual pace length. Remember, walk at a comfortable pace, neither trying to make the steps shorter or longer.

Don't pace alongside another student or else one of you will unknowingly set a pace identical to the partner. When you later work alone, the results will be incorrect.

When the pacing is completed, divide the measured distance (in inches or centimeters) by the number of paces to determine your personal **PL (pace length).**

Some mappers prefer to define the standard pace length as the distance determined by two steps; this type of pace would then be determined each time the student's right foot (or left, if you prefer) hits the ground (left-right-left). Others will be more comfortable in making the pace length the distance between each foot fall. A typical PL for an average sized person will generally fall between 2.2 to 2.5 feet (67–76 cm or 0.67–0.76 m). Shorter students will have smaller values while taller ones will have higher values. Two-step paces will be almost double in length.

Consistency in walking speed is the most important aspect because your pace length will tend to get shorter with a slower pace and longer when the speed is increased. It is imperative to concentrate on keeping a uniform speed!

Your instructor will provide a sheet outlining outdoor distances between known points to allow some practice in determining PL accuracy. Pace these routes and compare the resulting pace numbers to the predetermined values.

Part Two: Compass Directions and Bearings

The compass provided by your school may vary somewhat in appearance, but will have the features shown on Figure 7.5. The main part of the compass will be a flat platform (A) with a **turnable dial** or *compass housing* (B). One end of the platform will normally have a hole with a cord so that you can wear the compass around your neck, leaving your hands free for note-taking. One end of the platform will have a dark-colored **arrow** (C). This **direction of travel arrow** will be used for taking our *compass bearings.*

Always try to hold the compass so that the platform is as nearly level as possible so that the **magnetic compass needle** (D) can swing freely as the compass is turned. Note that the moveable *compass needle* has two labeled ends, generally a red or bright-colored end pointing to **magnetic north** and a white end pointing toward *magnetic south.* Your compass will also have an outline of an arrow engraved on the floor of the compass under the compass needle (*orienting arrow* [E]). This orienting arrow will turn when the compass housing is turned.

With the dark **platform arrow** (C) pointed away from you, point (aim) the compass at the wall clock in your classroom. Next, rotate the compass housing dial (B) so that the north end of the freely moveable compass needle (D) lies directly over the orienting arrow (E) engraved inside the compass housing. Note that the compass housing ring (B) has degrees and directions engraved on its surface. Read the bearing in degrees by noting where the compass housing numbers lie directly above the indicator arrow (C) on the compass base OUTSIDE the compass housing.

Check with your instructor to make sure you have the correct bearing, as the remainder of the exercise will depend on your being able to accurately determine bearings. (NOTE: There is a tendency for students to read the location of the north end of the magnetic compass needle.)

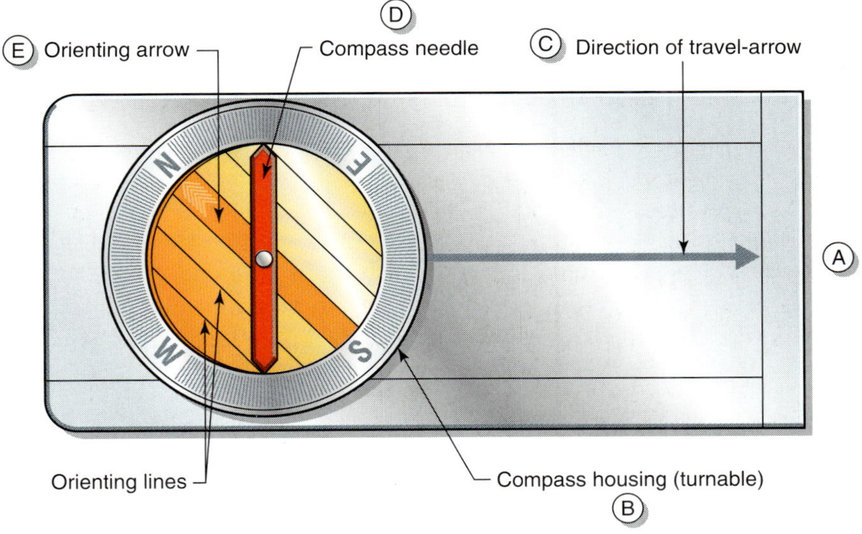

FIGURE 7.5 General features of compasses used for mapping and hiking.

Practice taking a few more bearings inside the classroom and confirming your results with the instructor. When you feel comfortable with these experiments, you are ready to move outside and try your new skills. Your instructor will provide you with a sheet of directions for taking bearings "to and from" specific points on your campus. Head outside and see how well you have learned the bearing part of this exercise and then come back to the classroom to compare values with those determined by your instructor.

Part Three: The Pace and Compass Project

On this campus, light poles and fire hydrants are identified by numbers (for example, on our campus, light posts are marked with either a letter or a letter and number designation, while fire hydrants only bear numbers). Perhaps your campus will allow you to put similar labels on identifiable features (light posts, hydrants, distinctive trees, etc.). Your instructor will provide a box containing slips of paper with numbers and/or letters, each slip indicating an identifiable campus feature.

Scout out the locations of your assigned features to make sure you know generally where everything is located. When ready, stand with your back at the starting feature (possibly a light post or a convenient tree) designated by your instructor.

If you work in teams, be sure to trade duties often so that each student learns the techniques. Start at a point selected by your instructor. One person should be pacing and counting the paces while another records the data and prepares a sketch map of the study area. Your map area should ideally include five or more items to include in your map. During the mapping, make a sketch of the area recording positions of trees, walks, major buildings, etc.

Using the instructions above, determine the compass bearing to the first item on your assigned site list. Record the bearing on your data sheet and carefully pace the distance between the two features. Record the pace number on the data sheet.

Stand with your back against the new feature (tree or light post) and determine the bearing to the next site on your list. Record the new bearing, pace off the distance to that feature, and record the pace numbers.

Repeat these steps for however many sites your instructor has assigned, making the last bearing pointing back to the starting point. Be very careful at each step of the process.

Part Four: Plotting the Data

We now need to see if you can produce a map from the data collected during the pace and compass part of this exercise. Before starting to plot your data, multiply each of your pace numbers between features by the individual pace lengths (PL) to convert the pace distances to feet or centimeters (or meters).

Your instructor will provide sheets of graph paper with ten squares to the inch. We will be using a map scale of one inch equals one hundred feet; therefore, one of the small squares on the paper equals ten feet on the ground. The longer lines (top to bottom) on the graph paper will become the north-south directions and the right to left lines will be the east-west directions.

Checking your measured distances, carefully determine where to begin your map.

INSIGHTS

Be very careful in recording the bearings or azimuths between locations. Errors can build up quickly, and your lines will not "close." In practice, you will seldom come back precisely to the point where you started, but, if you are significantly off, you will need to repeat the process and locate your error.

CAUTION

Watch out for iron! If you have a metal belt buckle made of iron, the compass needle may point to you. Likewise, do not stand directly against metal objects such as railings or lamp posts because the compass may give an incorrect reading.

INSIGHTS

If you are not careful at this step, your map may run off the paper. Your instructor will help confirm that you have chosen a reasonable starting point.

Use a standard circular protractor by placing the center point of the protractor on the starting point of your map, making sure that the 0° (360°) point on the protractor rim is pointing to the top of the sheet of graph paper. It will help visualize the process if the starting point is also located ON one of the north-south lines.

Note the first compass bearing you determined. If it was, for example, 60°, you would place a pencil dot on a point outside the protractor rim where the dial reads 60. With a ruler calibrated for the system you are using (inches or centimeters), determine the length of a line between your starting point and the location of the first pace and compass location outside.

Once you have measured this distance, draw a line on the map between your starting point and feature one. For instance, if the distance between your starting point and feature number one was 288 feet, using a scale of one inch equals 100 feet, you will need to draw a line 2.88 feet long (2.9 inches would probably be close enough). Make a clearly visible dot at the end of this line, draw a small circle around the far end of this first line, and label it with some letter or number that indicates site one on your map.

Move the protractor starting point so that it is now located over the dot at the end of line one. Carefully orient the protractor so that north is correctly indicated (0° or 360° is toward the top of the paper and along a north-south line). Now mark the bearing to your location three, plot the new direction, and mark the line length in the same way you determined the first site. Again, place a clearly evident dot at the end of the new line, circle it, and label the spot.

Continue plotting bearings and determining the new site locations until you have completed the series. The excitement comes when you determine the very last bearing and measurement. Did you end up exactly at your starting point? If so, congratulations! If not, what happened?

Perhaps you were not successful this first time. If time is available, go back outside and retake the bearings to find where you made your error. The last step where you hopefully ended up exactly at your starting point is called "closing the loop" and is the goal of every surveyor or mapper.

It's not likely that you will close exactly, so don't get too frustrated, but if you are very far off, you need to look for errors. Your instructor will tell you how much error is allowable.

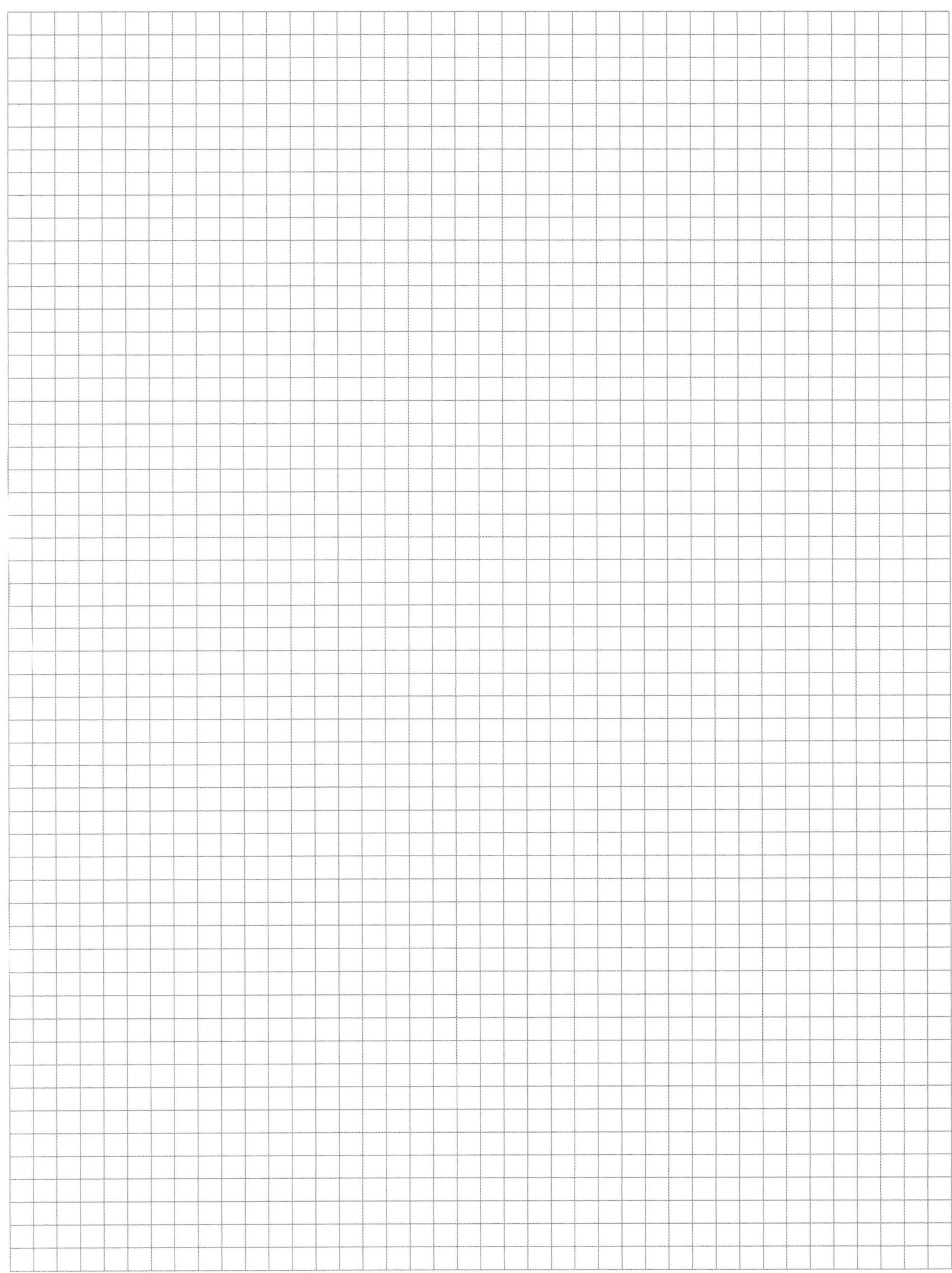

Geology Lessons Learned from Maps

Throughout this laboratory manual, and throughout your coursework in geology to date, you have been learning about geologic principles, materials, methods of investigation, and the information needed to interpret a large suite of geologic settings.

It is time to see if you can actually use what you have learned to interpret geology directly from common topographic maps. The following short discussions will each be focused on a topographic map that demonstrates some set of geologic data. Locate the points and answer each of the questions.

Consider the following as you work through this chapter. Some of the questions are stated in the form of thought questions. There sometimes will not be a perfect answer, but use the information you and your classmates have accumulated and see how close to the mark you can come.

The physical constraints of publishing forces this manual to use pieces of the entire maps; the originals are just too large and too expensive for every manual to contain them at full scale. Hopefully your school geology laboratory will have many, if not all, of these fine maps. If so, use the full-scale versions in order to get a better sense of the scenes you are analyzing.

The only order used in the presentation on the following pages was to group the maps in general categories.

■ ARID REGIONS

Antelope Peak, Arizona (15 minute series; 1:62,500 presented in a normal contour version and a shaded relief version)

1. The higher area in the southwestern portion of this map is a feature named Table Top Mountains. Note the following:

   a. The contour lines near the southwest corner of the map are very close together. What does this mean?

   b. The contour lines from Indian Butte to the northeastern corner of the map become wider and wider apart. Explain why this occurs.

   c. Each contour line is drawn as a saw-toothed line. What type of landscape feature do these jagged lines demonstrate?

   d. There are no evident rivers in the map area, but there are abundant signs of erosion on the map. Discuss the types of erosion and rainfall patterns responsible for this landscape.

2. The Antelope Peak, Arizona quadrangle map is presented here in two forms: normal contour version and a shaded relief version. In the shaded relief map, an artist has used the topographic details and a fixed sunlight angle to cast artificial shadows. The result is a realistic rendering of the landscape that is easier for some people to understand. Which do YOU like better and why?

3. Note that in the shaded relief version, the origin of that odd channel originating in the southwest part of the map becomes a great deal more evident.

4. If you have the full map at hand, what is the magnetic declination for this quadrangle map? How would you use this information to correct your compass readings? What is the scale of this map? (NOTE: There can be a number of different answers, all equally correct.)

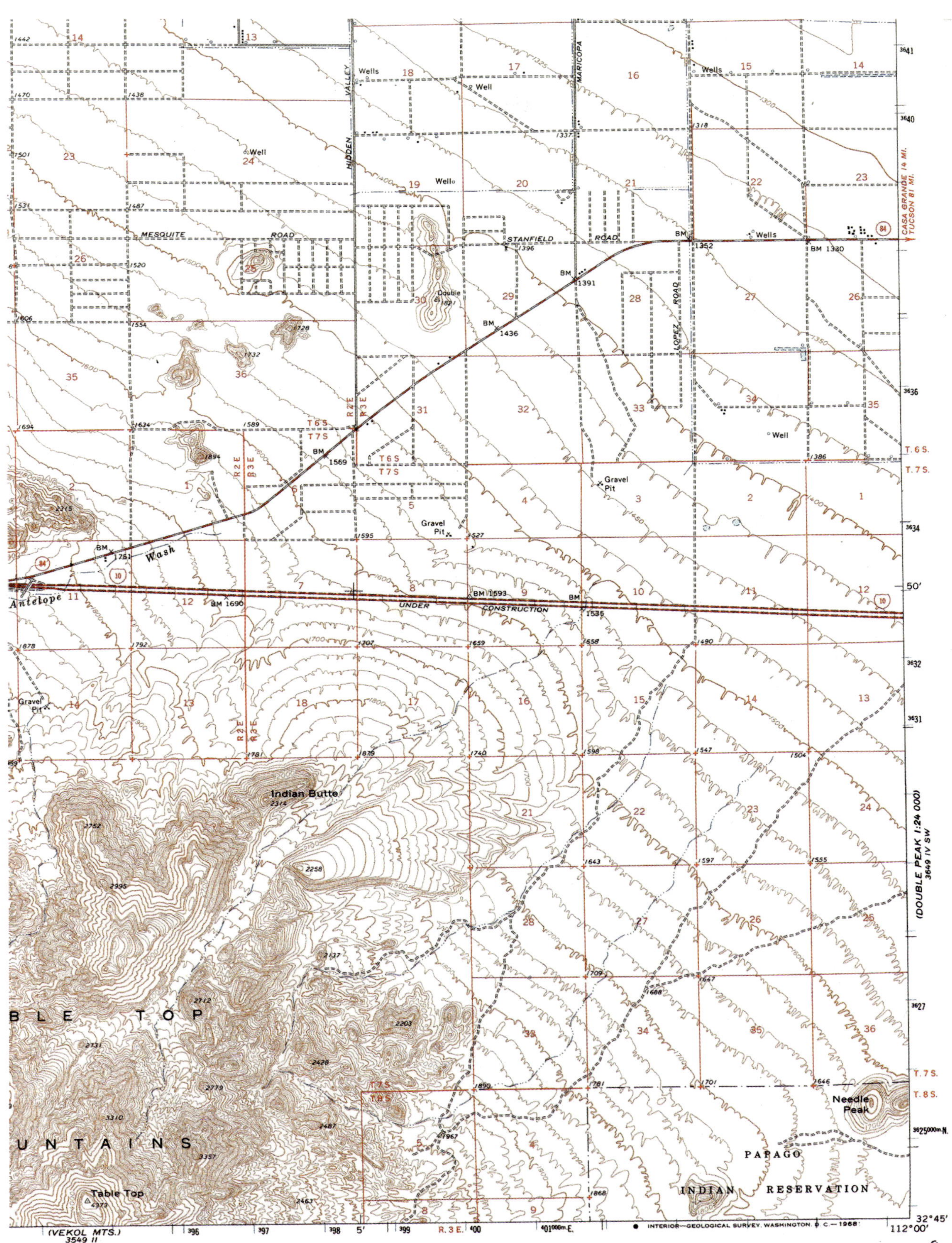

Antelope Peak, Arizona

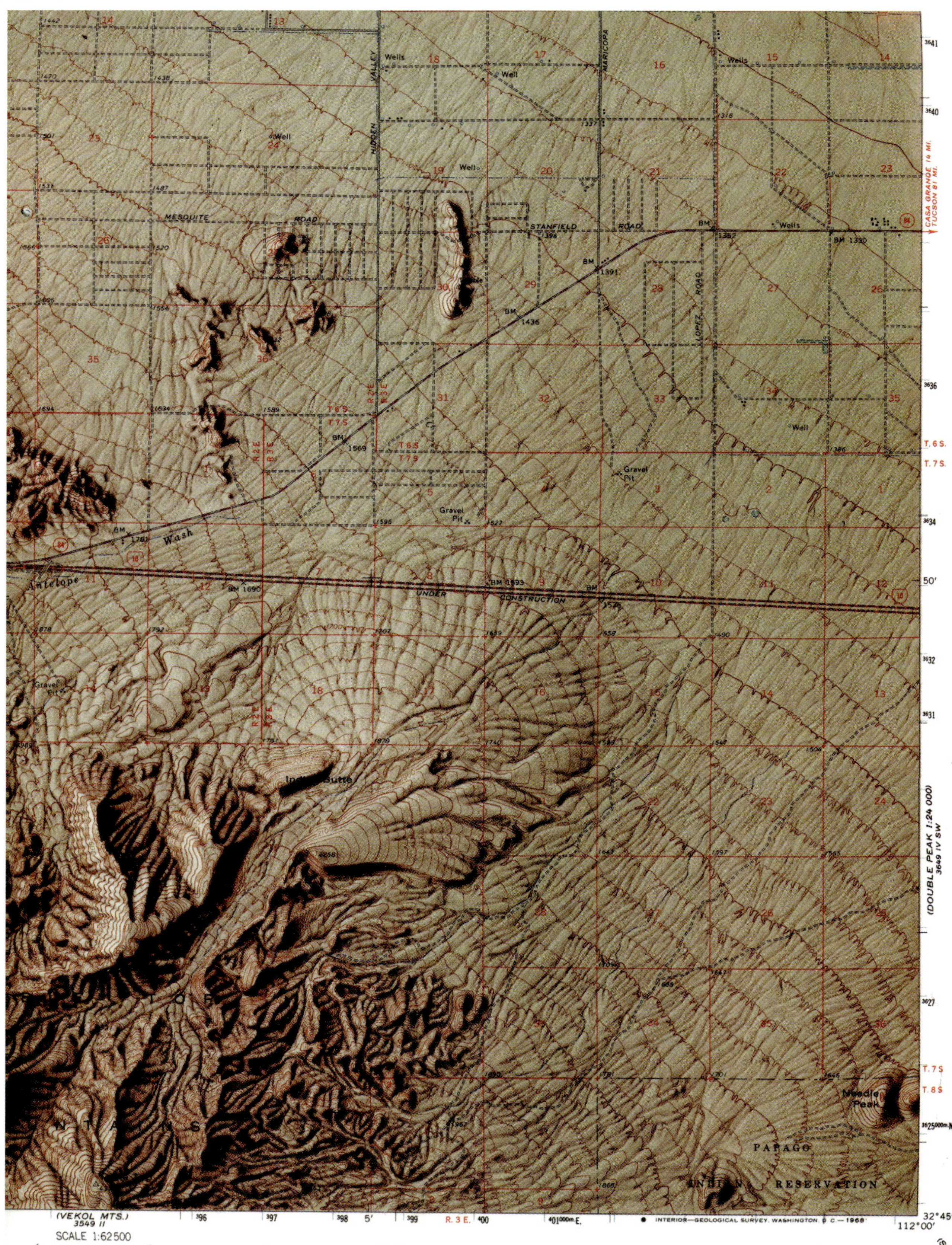

Antelope Peak, Arizona
Shaded Relief

Moab, Utah (15 minute series; 1:62,500)

This spectacular landscape contains the setting for the city of Moab, Utah, plus parts of the Arches National Monument and the edge of the Canyonlands National Monument, two of America's grandest national parks.

1. There is a schoolhouse indicated in the shaded area of Moab, Utah. Locate this schoolhouse to the nearest quarter-quarter-quarter using the PLS System (Township and Range System).

2. Long straight lines are rare in natural landscapes, and when they do occur, they are generally the result of some sort of tectonic activity, such as faulting. Check out the location of Moab and note that the town is located on the floor of a flat valley, with steep walls on either side. What sort of structural feature is formed where there are two facing normal faults and the area between has been down-dropped?

3. Note that the southwest part of the map between the Colorado River and the Moab canyon wall is marked by numerous small east-west trending shallow canyons. Any suggestions as to their origin?

4. The Colorado River makes a series of broad loops through this map area (you can see one of these on the west side of this map). Note that the river is in the bottom of a very steep-walled canyon. How did it dig this deep a canyon without meandering?

A related question involves a feature in the southwest corner of the map that is labeled Jackson Hole. It is associated with the canyon of the Colorado River. What processes produced this oddly semicircular feature with the tall central spire?

Moab, Utah

■ COASTAL LANDSCAPES

Aransas Pass, Texas (15 minute series; 1:62,500)

1. The tiny village of Port Aransas is a popular tourist area located across Corpus Christi Bay from the city of Corpus Christi, Texas. The long northeast to southwest trending feature is part of the barrier island complex that runs nearly the entire length of the Texas coastline. To the northeast it is known as Saint Joseph Island and southwest of Aransas Pass it bears the name of Mustang Island. Landward (behind the islands) lie a series of shallow water areas called lagoons (Aransas Bay, Marsh Bay, Corpus Christi Bay). The prevailing winds are from the east.

   a. Why are Saint Joseph Island and Mustang Island both so narrow? (Think about sea levels).

   b. Why are there a series of low lying islands and large sandy shoal areas on the landward side of the island and Aransas Pass?

   c. What processes produced these large sandy shoals?

   d. Notice that Saint Joseph Island has a wider end near Aransas Pass and another wider end at the northeast edge of the map, but in between, the island is very narrow. However, there is a broad sand flat west of the island. What coastal weather and marine event could have produced this odd island shape?

   e. Note from the contour data that there are high areas immediately west of the beaches, but that the areas further west become very flat, very low-lying, and have an irregular margin compared to the eastern edges of the islands. Account for this east to west change in the character and profile of the islands.

2. This map has the normal contour data for the land surfaces, and also contains much contour data that refers to the seafloors. Remember, the seafloor contour data are in numbers that grow larger as the water gets deeper, that is, the reverse of land contour data.

   a. Scan the depth information. What does it tell you about the Texas coastline?

   b. If you were the captain of the new ocean liner, the Queen Mary II, would you want to bring it into Corpus Christi Bay through Aransas Pass? Why or why not?

   Queen Mary II specifications

Length	345 metres/1132 feet
Beam	41 metres/135 feet
Draft	10 metres/32 feet 10 inches
Gross Tonnage	150,000 tons
Passengers	2,620; Crew 1,253
Cost (Est.)	$800 million dollars

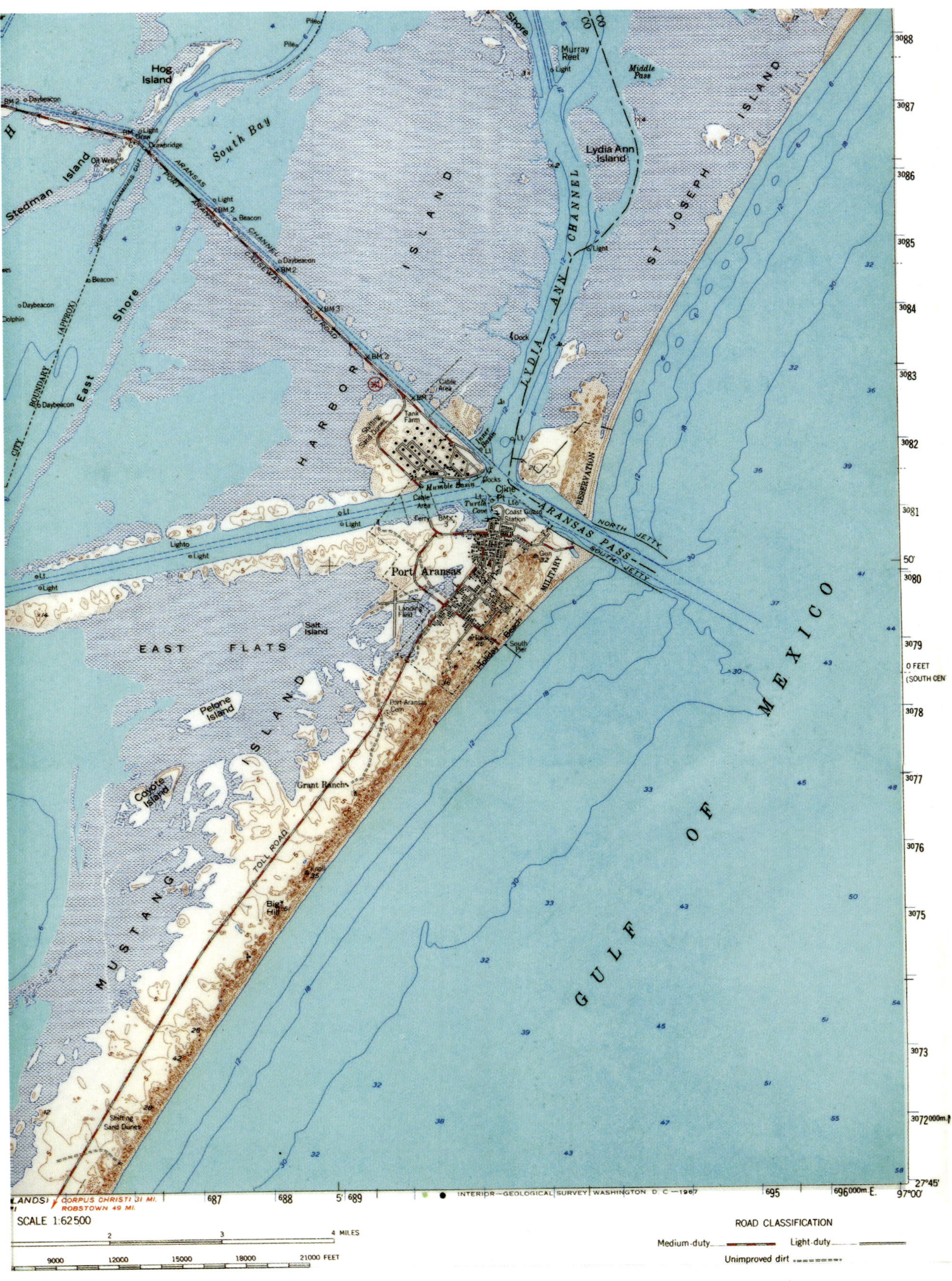

Aransas Pass, Texas

Cayucos, California (15 minute series; 1:62,500)

1. This area north of Los Angeles shows evidence of a strong longshore current system and strong oceanic waves. Can you name at least two evident features to support this statement?

2. A prominent feature, Morro Rock, juts out into the sea west of the town of Morro Bay. What is this feature and the sand bank that connects it to the shore line called?

3. Given the effects of the longshore current system, what name should be applied to the sandy beach that forms the seaward margin of Morro Bay?

4. As noted above for the Aransas Pass map, this map contains contour data that refer to the seafloors. Compare the subsea contour data for the Morro Bay region with the same data from Aransas Pass.

    a. How do the two coastline offshore areas differ?

    b. What large-scale tectonic processes produce the two different coastlines?

5. Why is the back of Morro Bay (marked State Park) being filled in so rapidly near the mouth of Chorro Creek?

6. How large a boat can you use in Morro Bay. (To answer this, determine the depth of the water.)

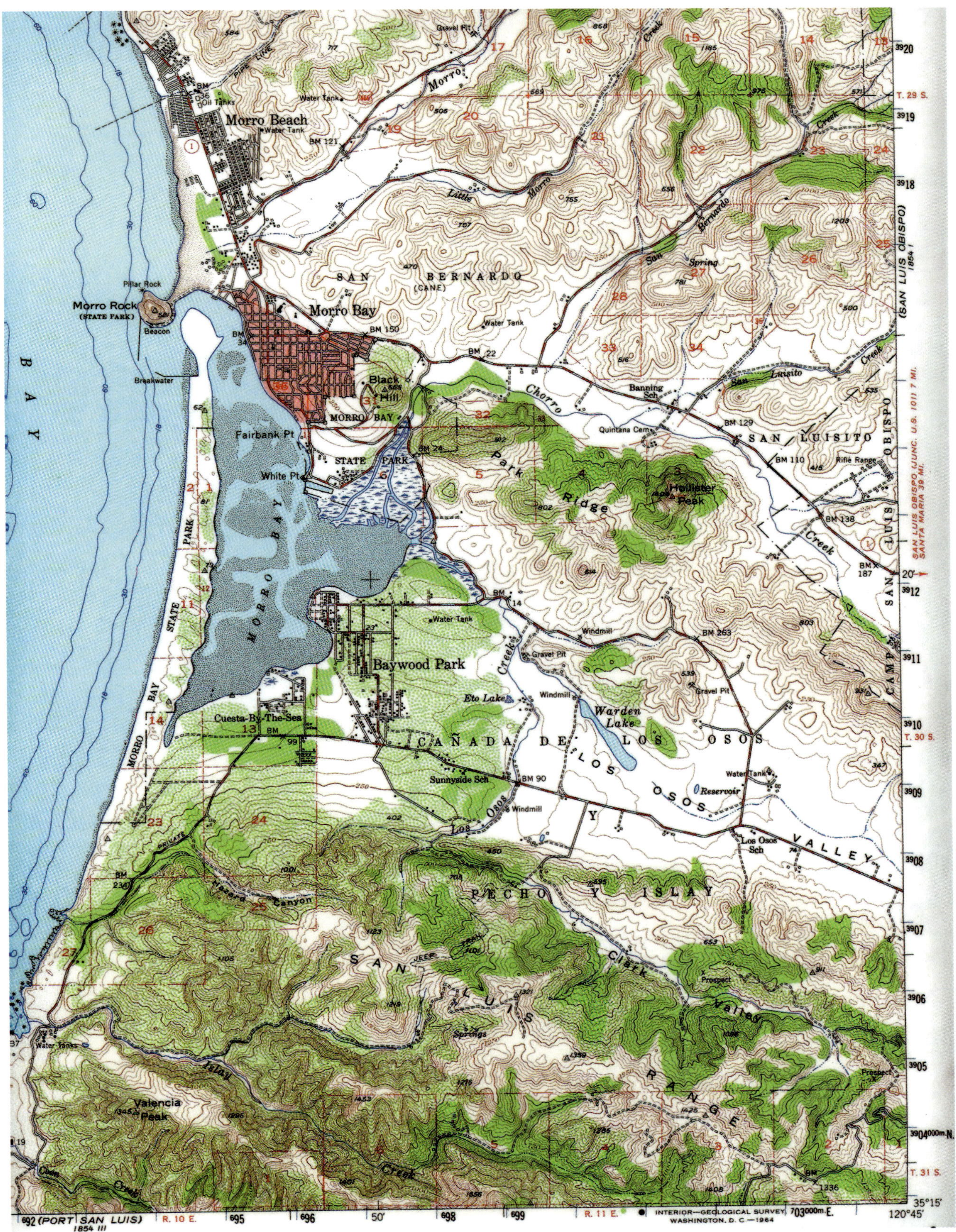

Cayucos, California

Point Reyes, California (15 minute series; 1:62,500)

1. As mentioned earlier, long straight lines are rare in nature unless produced by tectonic activity, as is the case in this map. What feature produced the elongated shape and straight sides of Tomales Bay?

2. Using the contour data, note the difference in the slopes on the east and west edges of Tomales Bay. What might account for this? (NOTE: This is related to question one.) HINT: The slopes on the east side are rolling hills developed on shales, mudstones, and decomposed metamorphic rocks, while the west margin of the bay is almost entirely composed of igneous rock.

3. What do all the little stars around the end of Tomales Point mean? NOTE: The stars are only located on the coastline north of Kehoe Ranch. How do the coastlines north and south of this point differ?

Point Reyes, California

■ TECTONICALLY ACTIVE AREAS

Harrisburg, Pennsylvania (15 minute series; 1:62,500)

1. What large-scale tectonic event produced the strikingly different landscapes northwest and southeast of Blue Mountain?

2. Given what you have learned about the geologic history of the United States east coast region, when did this tectonic event occur?

3. Note that the mountains are named (from southeast toward the northwest): Blue Mountain, Second Mountain, Third Mountain, and Peales Mountain (very top of the map). What probable structure underlies this region (anticlines, synclines, faults, etc.)?

4. The Susquehanna River cuts through the center of the map. In fact, it cuts right through each of the northeast-southwest trending mountain ridges. How did it do this?

5. Consider how this landscape would look to you if you were a pioneer in the late 1700s planning to head west.

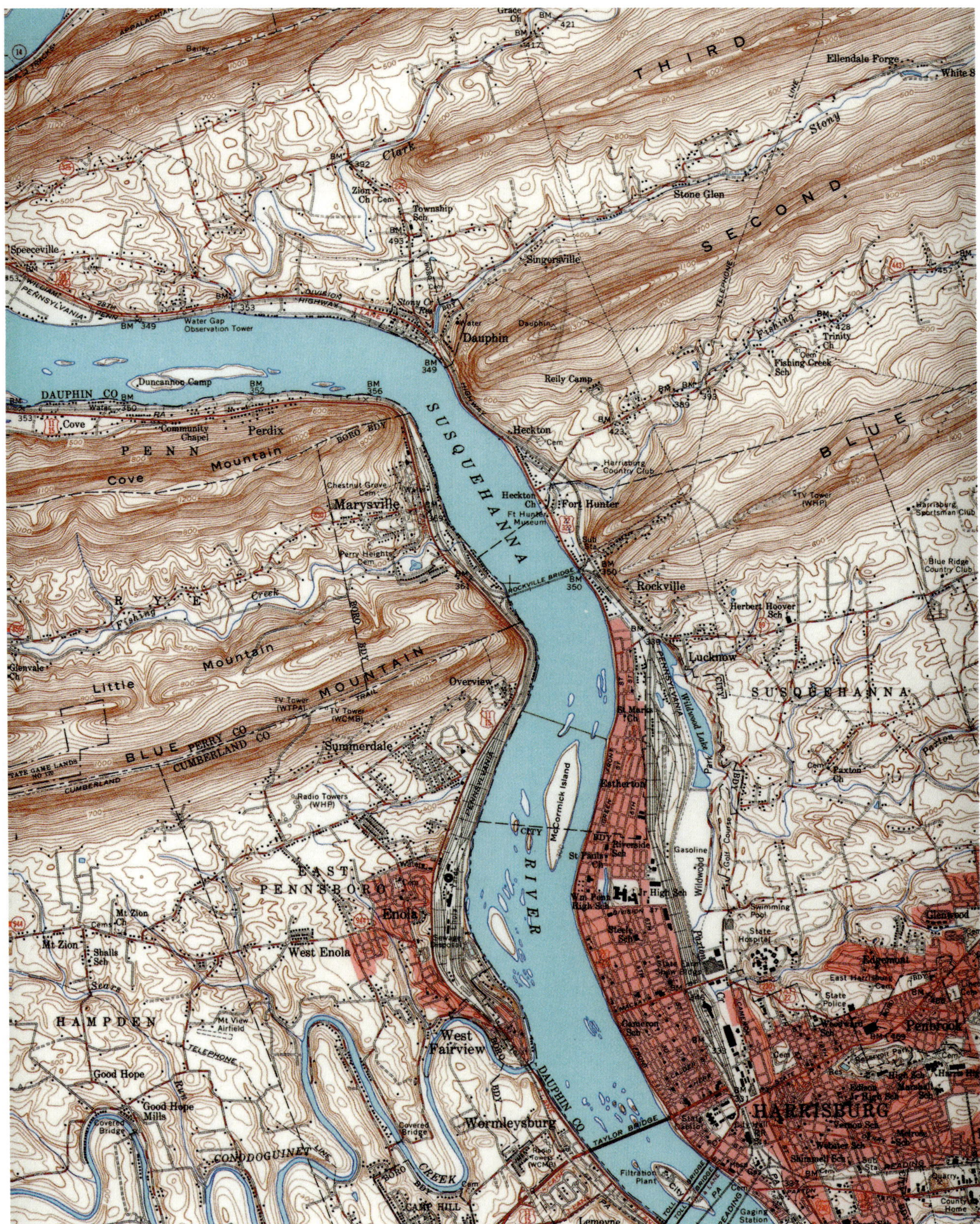

Harrisburg, Pennslyvania

Strasburg, Virginia (15 minute series; 1:62,500)

1. This landscape looks much like the one we saw on the Harrisburg, Pennsylvania map, with several striking differences. Note the patterns formed by the mountain ridges and the stream channels.

    a. Massanutten Mountain and Green Mountain parallel Passage Creek, and the creek channel bounces back and forth between the two ridges. In the same manner, the Shenandoah river is contained between Massanutten Mountain and the higher land to the southeast. What is the name for this type of stream drainage?

    b. Why are the streams apparently trapped between the mountains, when in the Harrisburg area the streams cut directly across the mountain ridges?

2. Given what you have learned about the geologic history of the United States east coast region, when did this tectonic event occur? Was this the same event that produced the landscape seen at Harrisburg, Pennsylvania?

Strasburg, Virginia

■ VOLCANIC LANDSCAPES

Menan Buttes, Idaho (7.5 minute series, 1:24,000)

1. The Menan Buttes are classic volcanic cones. From their shapes, sizes, and height, what type of volcano are they? (Consider shield volcanoes, Cinder Cone volcanoes, and stratovolcanoes).

2. Examine the contour lines both outside the cones and inside the craters. Do you see any differences?

3. How many episodes of volcanic activity seemed to have occurred in this map area?

4. How do the elevation markings differ between: X = 5035 (in the crater of the southern cone); the triangle with the elevation 5619 on the north rim of the northern cone, and the spot to the west of the two cones where there is an elevation of 4826 and the abbreviation "BM"?

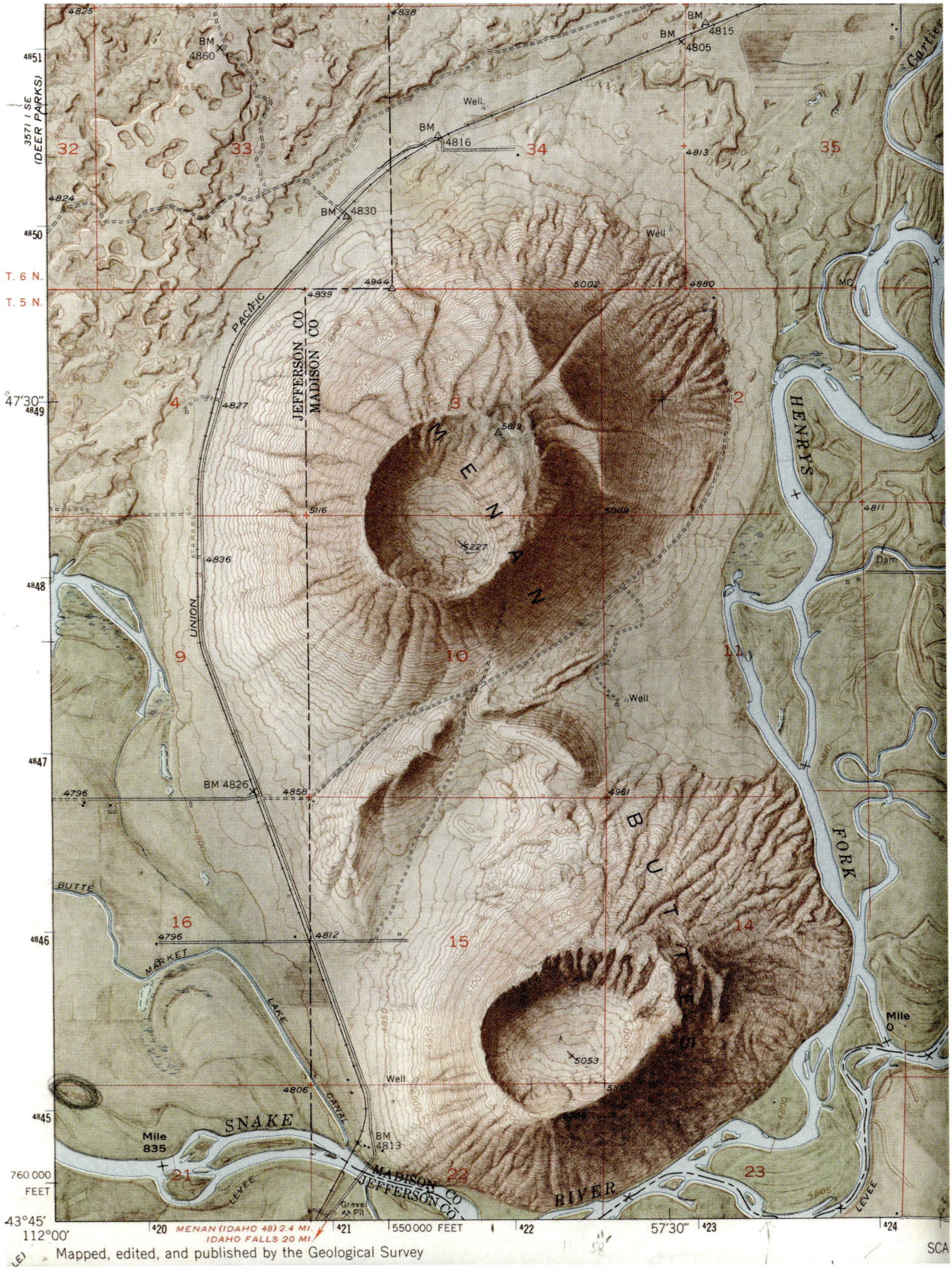

Menan Buttes, Idaho (7.5 minute series, 1:24,000)

Ship Rock, New Mexico (15 minute series; 1:62,500)

1. The feature known as Ship Rock may well be illustrated in your geology textbook. This spire of rock rises well over 600 meters high and 500 meters in diameter straight out of the desert floor. It is the remains of an ancient volcanic structure, but one that has been so severely eroded that the cone is now absent. What we see when we visit this spot is the remnants of the neck, or conduit, that once carried magma to the surface.

   Look at the contour data for the main peak relative to the surrounding area.

2. Because the sides of Ship Rock are nearly vertical for 300 meters or more, it has been a major target for rock climbers. The site, however, is one of the more sacred sites to the Navajo Indians, current owners of the site. Today Ship Rock is off-limits to ALL climbing. Is this justified?

3. What features within this mapped area tell you about the climate, drainage, vegetation, and utilization by the local inhabitants?

4. Some of the more interesting and prominent features of the Ship Rock site are the long, tall stone walls that radiate out away from the central volcanic neck (the main examples run west from the neck and south from the neck, although there are several more). What are these features and how did they form?

5. When you have answered question 4, consider how the features described there became exposed to form vertical rock walls nearly 30 meters tall.

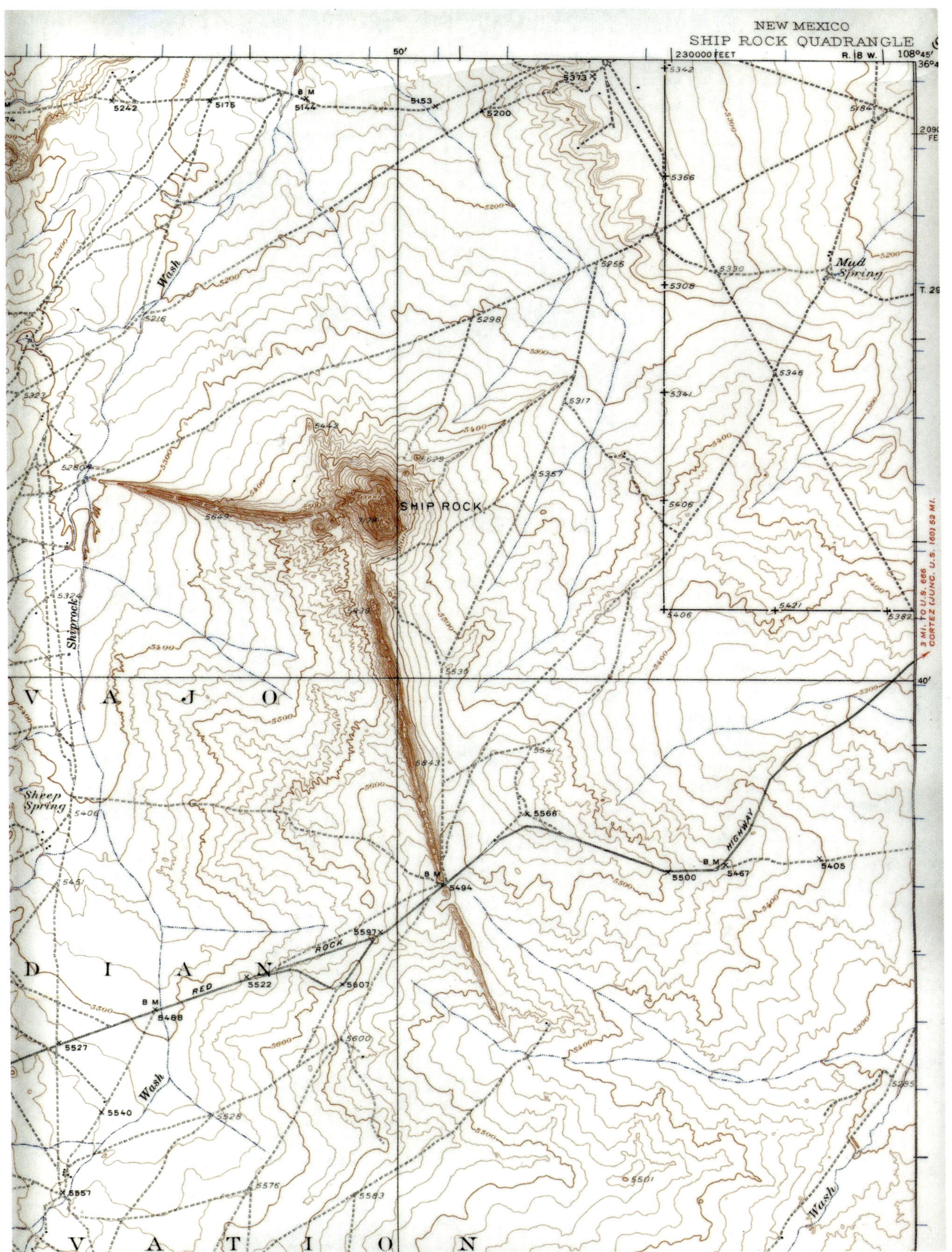

Ship Rock, New Mexico (15 minute series; 1:62,500)

Mount Rainier, Washington (15 minute series; 1:62,500)

1. This beautiful mountain is located along a volcanic trend termed the Cascade Mountains, running from northern California, and across Oregon and Washington. The Cascades are all volcanic and many of them are active or semi-active. Why did this chain form along the western edge of North America?

2. Mt. Saint Helens is also one of the Cascade volcanoes and it exploded with devastating force in 1980. One half of a cubic mile of debris was ejected from the volcano. Fifty-seven people lost their lives and the Columbia River temporarily became impassible to ocean-going vessels headed to Portland, Oregon. There were also 5,000 black-tailed deer, 1500 elk, 15 mountain goats, and 300 black bears who perished and this was in a very low human density area. Find Mt. Rainier on a map and suggest how you might feel if you were living in Seattle or Tacoma, Washington.

3. Mt. Rainier today is not believed to be dangerous, and the map shows some of the beauty of the mountain a tall, symmetrical, and often snow-covered cone.

   What is the elevation of the highest point of the mountain? What is the total relief between the top of the mountain and the Nisqually Park Entrance Ranger Station (southwest from the peak and a third of the way up the left map margin).

4. Noting the pattern of glacier development and stream channels on the mountain, what is this type of drainage pattern called?

Mount Rainier, Washington (15 minute series; 1:62,500)

GLACIAL TERRANE

Chief Mountain, Montana (Southeastern Map Area) (30 minute series; 1:125,000)

1. This area, within Glacier National Park, is heavily modified by the action of ancient and modern glaciers. Examine the area within this map section and see how many examples of the following glacial features you can identify:

   - Cirques

   - Arêtes

   - Horns

   - Hanging Valleys

   - U-Shaped Valleys

2. What is the tallest named peak in this mapped area? What is the shortest named peak?

Chief Mountain, Montana
Southeast

Chief Mountain, Montana (Northwestern Map Area) (30 minute series; 1:125,000)

1. Note the numerous small lakes in this area: Redhorn Lake, Lake Janet, Bench Lake, Sue Lake, and many others. Why are there so many of these tiny lakes and what processes produced them (be specific with your answers)?

2. Waterton Lake is located at the north edge of this map area. Using the modern stream drainages, how many glaciers once fed the main glacier that carved the valley now occupied by the lake? (NOTE: There can be a large variation in the answers. Be prepared to defend your total.)

3. If you have access to the whole map, answer these questions. What are the various scale types used to describe this map? Consider the quadrangle's general scale, the ratio scale, the bar scale, and the verbal scale.

Chief Mountain, Montana
Northwest

Holden, Washington (15 minute series; 1:62,500)

1. This area is much like that seen on the Chief Mountain, Montana sheet, and here again modern glaciers are nearly absent. Locate as many of the features that demonstrate the presence of now vanished glacial ice.

   - Cirques
   - Arêtes
   - Horns
   - Hanging Valleys
   - U-Shaped Valleys

2. Looking at the preserved patches of glacial ice, where are they located in the landscape and why have they persisted when the ice everywhere else is now melted?

3. If you were given the job of making contour maps in this area and the customer stipulated that he wishes the maps to use contour intervals of 5 feet, is this a good idea? What do YOU think would be the best contour interval and why?

4. What is the scale of this map? How did the map scale change affect your ability to locate and describe the glacial terrane features?

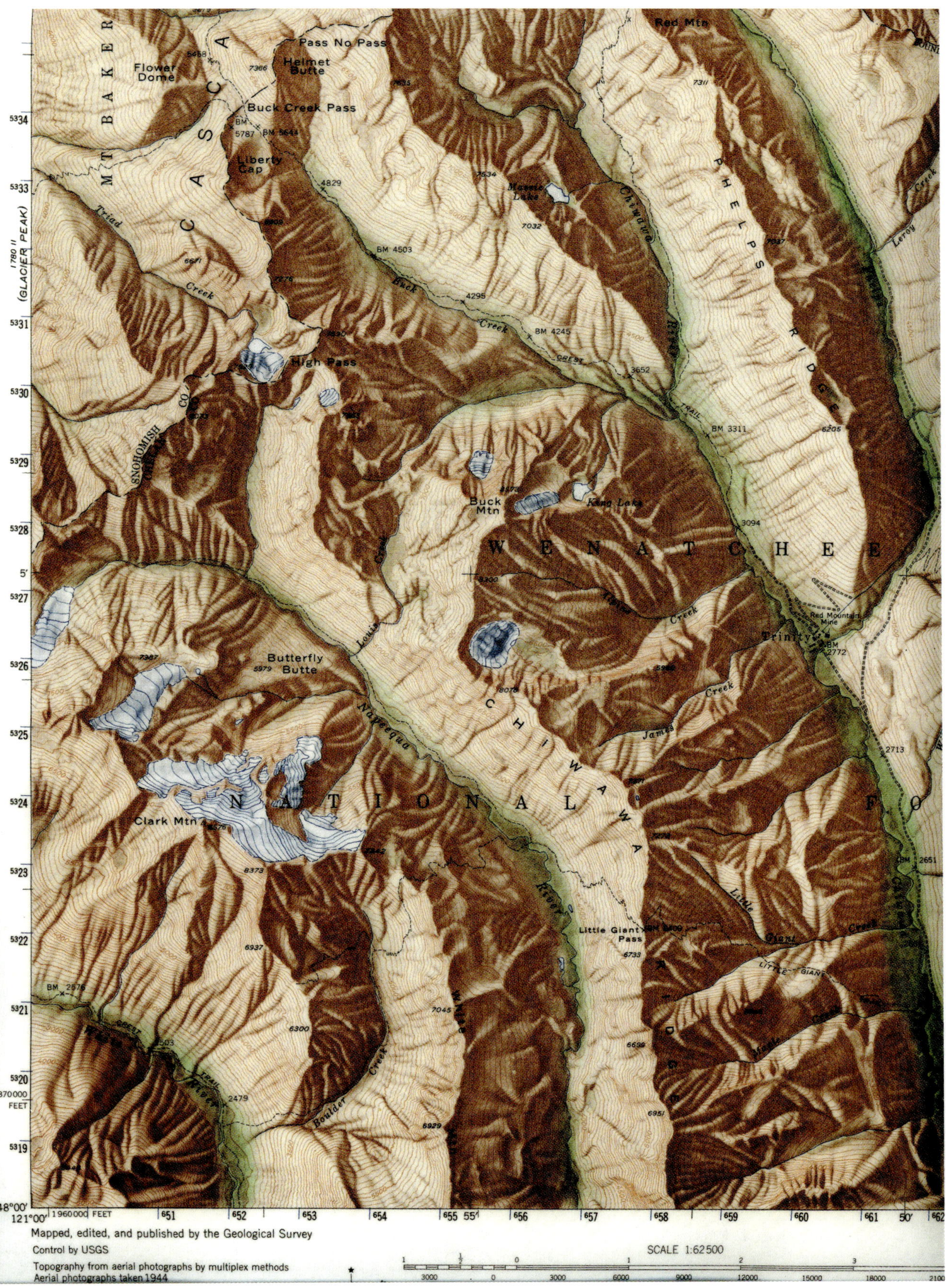

Holden, Washington

Whitewater, Wisconsin (Southeastern Map Area) (15 minute series; 1:62,500)

1. This region was heavily modified by glacial activity in the Pleistocene, but the nearest mountains are hundreds of miles away. What type of glaciers was present here, in contrast to the glaciers we observed on the Holden, Washington; Chief Mountain, Montana; and Mount Rainier, Washington maps?

2. Kettle Moraine State Forest preserves one of the more obvious glacial features in the map area. Why is the park area so much higher in elevation than the surrounding area? And, why is the park area long and linear?

3. The lakes are scattered and more or less round. Given that this is an area where glaciers were active, explain how these lakes were probably formed.

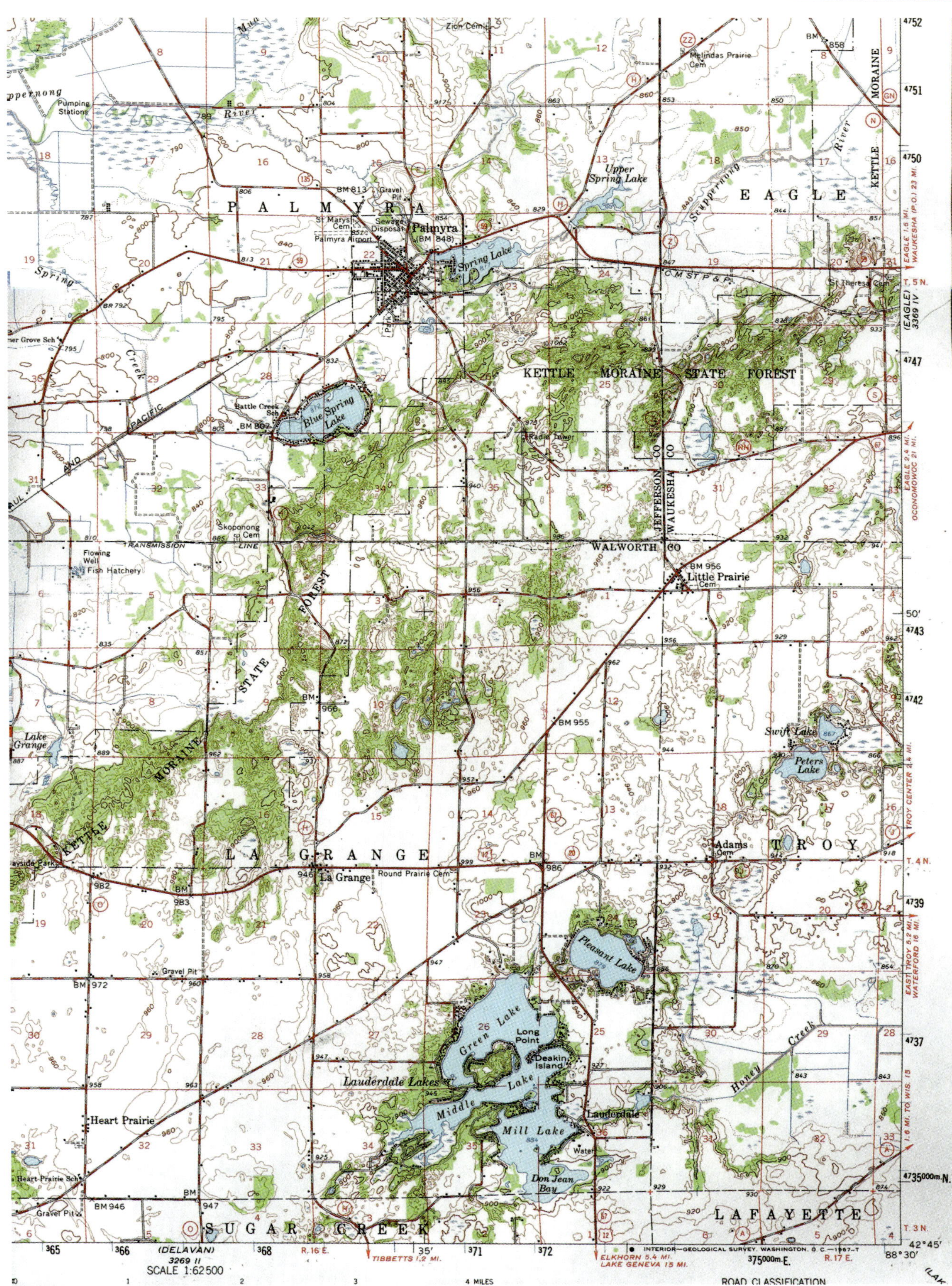

Whitewater, Wisconsin
Southeast

Whitewater, Wisconsin (Northwestern Map Area) (15 minute series; 1:62,500)

1. Streams in the area are relatively rare, and those that are present seem to have a hard time deciding whether to flow straight, meander widely or only slightly. Can you see something in this map area that explains the odd drainage setting?

2. Why do most of the roads on these two Whitewater maps run either north-south or east-west?

3. What is the longitude and latitude of the northwest corner of the map?

4. What is the location to the nearest quarter-quarter-quarter in the PLS System of the gravel pits near the McLery Cemetery? What symbol tells you precisely where the gravel pits are located?

5. Where is the highest point in this mapped area? Where is the lowest point? What is the total relief between them?

Whitewater, Wisconsin
Northwest

■ MISCELLANEOUS FEATURES

Refuge, Arkansas-Mississippi (15 minute series; 1:62,500)

1. This area, dominated by the channels of the Mississippi River, is a very complex one for mapmakers. Note the top of the map where the Range numbers appear. In the northeast above Miller Bend the map states R 1 E (Range 1 East). A short distance to the west, the same map margin states that the location in fact, R 10 W (Range 10 West). Similarly, along the west edge of the mapped area, it states that the locations are within township (tier) areas for T 14 S and T 15 S. On the opposite side of the map (not shown in this panel), the margin states that the area is within T 18 N and T 19 N.

   a. What is going on here to explain this discrepancy?

   b. Why did the mapmakers appear to use two very different systems on one map?

2. Note that each of the large island land areas seems to be in another state and the boundary lines between states seem to meander widely and with no apparent pattern.

   a. How was the boundary of the two states originally determined?

   b. How likely is it that the apparent boundary today is, in fact, the actual boundary formed when the states were founded? Why?

3. In Mark Twain's great book *Life on the Mississippi* he tells of the difficulties in becoming a riverboat pilot. He was required to remember the details of every river bend between New Orleans, Louisiana and Cairo, Illinois. This was vital, because the river was constantly changing and a pilot had to know where the deeper and shallower parts of the river were located in day or night. A pilot who failed would surely run aground.

   a. Locate places on this map that show major and minor shifts in the river channels.

   b. Can you identify projects that have been done to the river in an attempt to stabilize it today?

4. Many years ago, the author's family farmed an area of similar landscape along the Missouri River in Iowa. Their farm was in an area that resembled the Ashbrook Point area (northwest bend of the river). Over the years their farm grew larger and larger and cost them nothing. How could this have happened and on what side of the river was their farm located?

5. If you have access to the entire map, answer these questions. What is the scale of this map? Consider the quadrangle's general scale, the ratio scale, the bar scale, and the verbal scale.

Refuge, Arkansas-Mississippi

Interlachen, Florida (15 minute series; 1:62,500)

Located about 35 kilometers east of Gainesville, Florida and south of Jacksonville, this very strange landscape preserves many clues to its formation.

   a. Notice that water seems to be everywhere, but that streams are nearly absent.

   b. Areas without lakes tend to be covered with marsh and swamp.

   c. Check contour data and determine the local relief.

      1. How did all these lakes form?

      2. Why are there so few streams and rivers?

      3. Why do the lakes seem to be random in distribution?

      4. If you were to dig a deep hole anywhere in this mapped area, what type of bedrock would you encounter?

Interlachen, Florida

■ APPENDIX A—GEOLOGICAL TIMES SCALE AND LINKS

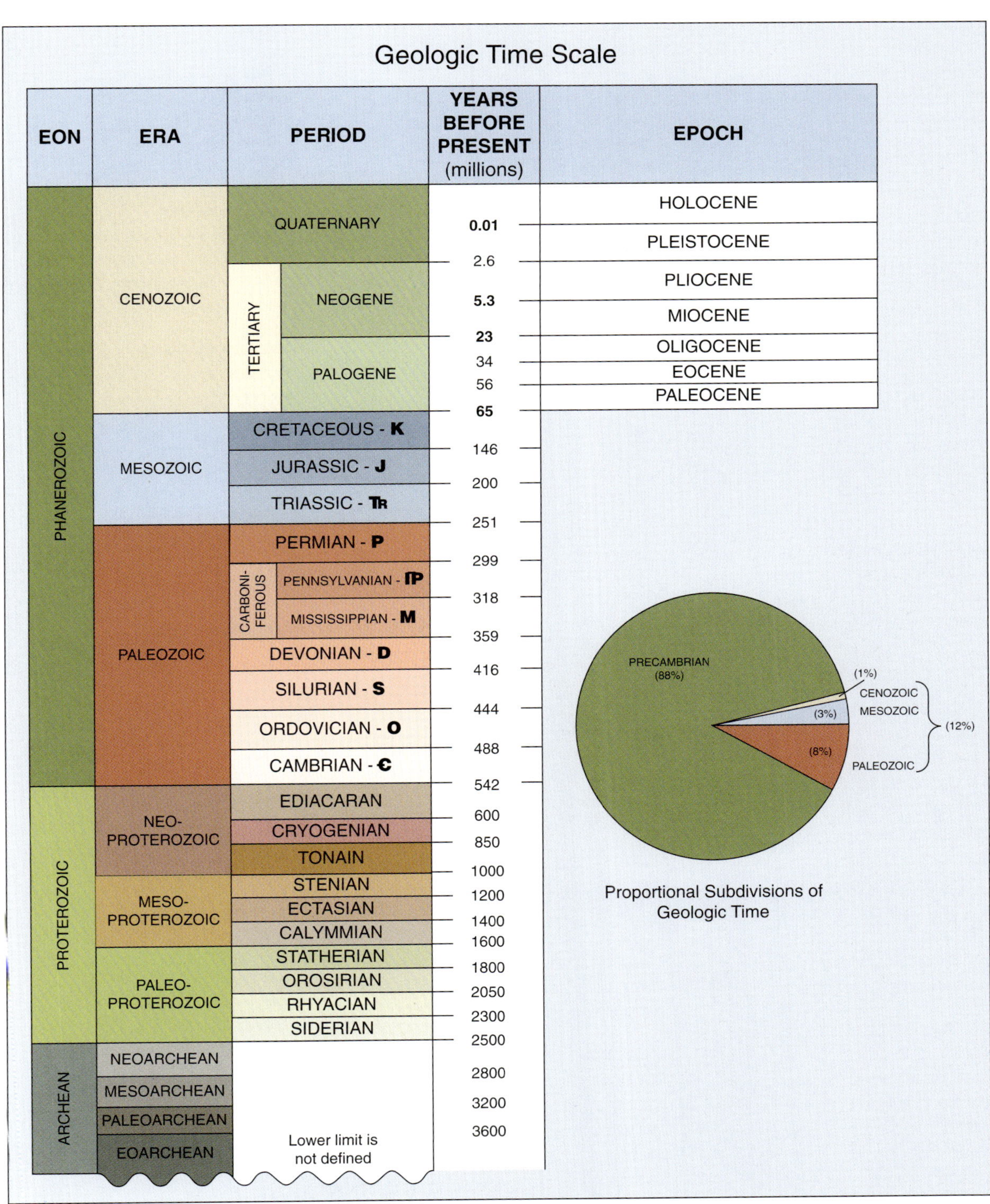

Adapted from the International Commission on Stratigraphy (ICS).

APPENDIX B—STANDARD U.S. GEOLOGICAL SURVEY MAP SYMBOLS

For more topographic map symbols go to **http://erg.usgs.gov/isb/pubs/booklets/symbols/**

Topographic Map Symbols

BOUNDARIES

National

State or territorial

County or equivalent

Civil township or equivalent

Incorporated-city or equivalent

Park, reservation, or monument

Small park

LAND SURVEY SYSTEMS

U.S. Public Land Survey System:

Township or range line

Location doubtful

Section line

Location doubtful

Found section corner; found closing corner

Witness corner; meander corner

Other land surveys:

Township or range line

Section line

Land grant or mining claim; monument

Fence line

ROADS AND RELATED FEATURES

Primary highway

Secondary highway

Light duty road

Unimproved road

Trail

Dual highway

Dual highway with median strip

Road under construction

Underpass; overpass

Bridge

Drawbridge

Tunnel

BUILDINGS AND RELATED FEATURES

Dwelling or place of employment: small; large

School; church

Barn, warehouse, etc.: small; large

House omission tint

Racetrack

Airport

Landing strip

Well (other than water); windmill

Water tank: small; large

Other tank: small; large

Covered reservoir

Gaging station

Landmark object

Campground; picnic area

Cemetery: small; large

RAILROADS AND RELATED FEATURES

Standard gauge single track; station

Standard gauge multiple track

Abandoned

Under construction

Narrow gauge single track

Narrow gauge multiple track

Railroad in street

Juxtaposition

Roundhouse and turntable

TRANSMISSION LINES AND PIPELINES

Power transmission line: pole; tower

Telephone or telegraph line

Aboveground oil or gas pipeline

Underground oil or gas pipeline

CONTOURS

Topographic:

Intermediate

Index

Supplementary

Depression

Cut; fill

Bathymetric:

Intermediate

Index

Primary

Index Primary

Supplementary

MINES AND CAVES

Quarry or open pit mine

Gravel, sand, clay, or borrow pit

Mine tunnel or cave entrance

Prospect; mine shaft

Mine dump

Tailings

SURFACE FEATURES

Levee

Sand or mud area, dunes, or shifting sand

Intricate surface area

Gravel beach or glacial moraine

Tailings pond

VEGETATION

Woods

Scrub

Orchard

Vineyard

Mangrove

COASTAL FEATURES

Foreshore flat

Rock or coral reef

Rock bare or awash

Group of rocks bare or awash

Exposed wreck

Depth curve; sounding

Breakwater, pier, jetty, or wharf

Seawall

BATHYMETRIC FEATURES

Area exposed at mean low tide; sounding datum

Channel

Offshore oil or gas: well; platform

Sunken rock

RIVERS, LAKES, AND CANALS

Intermittent stream

Intermittent river

Disappearing stream

Perennial stream

Perennial river

Small falls; small rapids

Large falls; large rapids

Masonry dam

Dam with lock

Dam carrying road

Intermittent lake or pond

Dry lake

Narrow wash

Wide wash

Canal, flume, or aqueduct with lock

Elevated aqueduct, flume, or conduit

Aqueduct tunnel

Water well; spring or seep

GLACIERS AND PERMANENT SNOWFIELDS

Contours and limits

Form lines

SUBMERGED AREAS AND BOGS

Marsh or swamp

Submerged marsh or swamp

Wooded marsh or swamp

Submerged wooded marsh or swamp

Rice field

Land subject to inundation

APPENDIX C—GRAPHICAL PATTERNS FOR COMMON ROCK TYPES

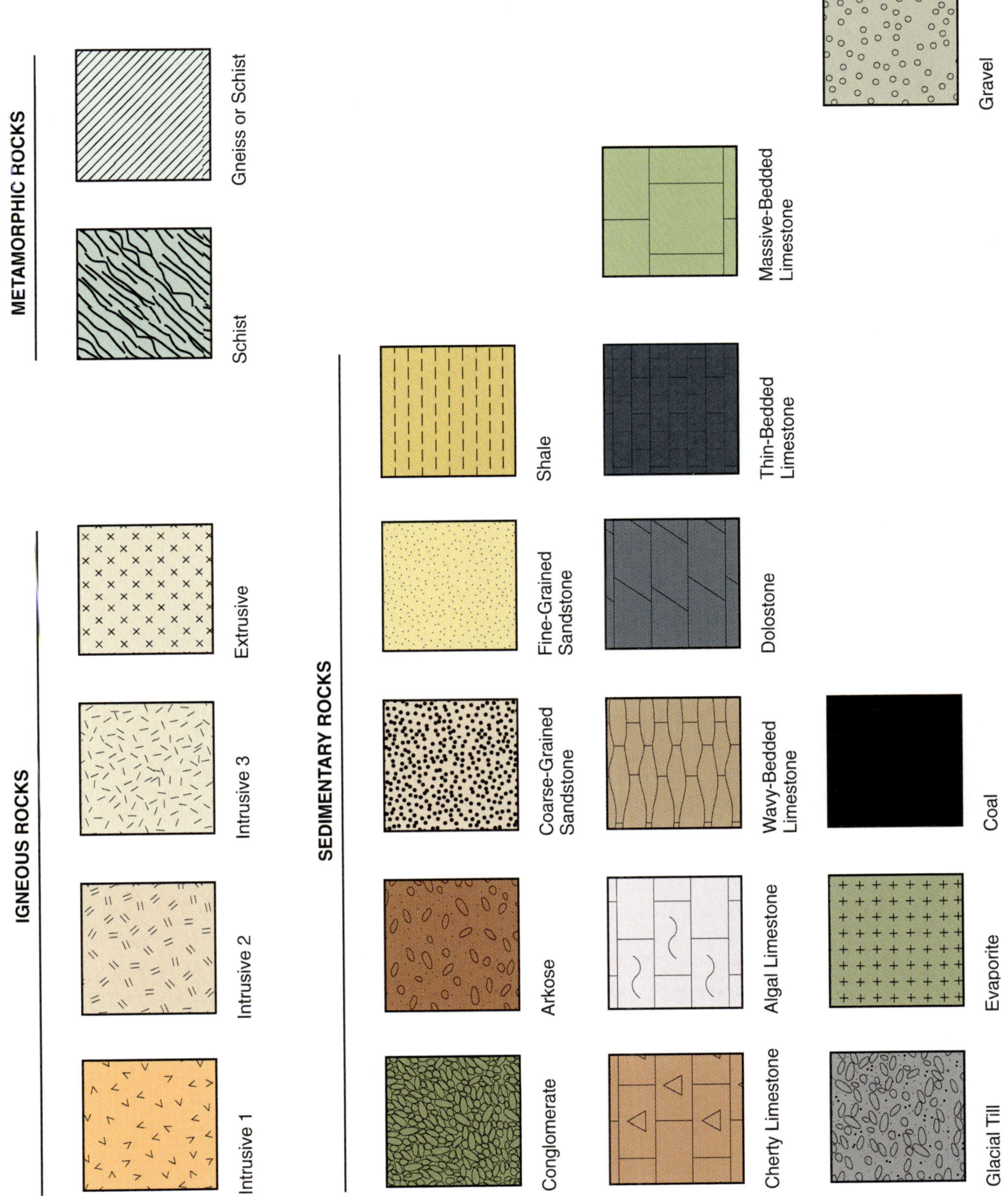

■ APPENDIX D—ENGLISH AND METRIC UNITS OF MEASUREMENT

Linear Measurements

12 inches = 1 foot
3 feet = 1 yard
5,280 feet = 1 mile

1000 microns = 1 millimeter (mm)
10 millimeters = 1 centimeter (cm)
10 centimeters = 1 decimeter (dm)
10 decimeters = 1 meter (m)
100 centimeters = 1 meter
1000 millimeters = 1 meter
1000 meters = 1 kilometer (km)

1 inch = 25.4 millimeters
1 inch = 2.54 centimeters
1 foot = 30.48 centimeters
1 foot = 3.048 decimeters
1 foot = 0.3048 meter
1 yard = 0.9144 meter
1 mile = 1.6093 kilometer

1 millimeter = 0.04 inch
1 centimeter = 0.4 inch
1 decimeter = 3.94 inches
1 meter = 39.4 inches
1 meter = 3.28 feet
1 meter = 1.094 yards
1 kilometer = 3280.8 feet

1 kilometer = 0.62 mile

Measurements Used by Land Surveyors

Distance/Length

1 link = 7.92 inches
1 rod = 25 links = 16.5 feet = 1/4 chain
1 chain = 100 links = 66 feet = 4 rods
1 furlong = 660 feet = 40 rods
1 square acre = 208 3/4 feet on a side
1 statute mile = 5280 feet = 1.609 kilometers = 320 rods = 80 chains = 8 furlongs
1 Nautical Mile = 1000 meters = 0.6214 mile (approximately 5/8 mile)
1 meter = 100 centimeters = 39.37 inches = 3.2808 feet
1 vara (used in measuring Spanish Land Grants) = 33 1/3 inches
36 varas = 100 feet
1 arpent (French Land Grants) = length of one edge of a square arpent = about 11.5 rods

Area

1 acre = 45,560 square feet = 160 square rods = 10 square chains
1 arpent = 0.85 acre
1 township = 36 square miles = 36 sections = 23,040 acres
1 section = 1 square mile = 640 acres
1/4 section = 160 acres
1/4, 1/4 section = 40 acres
1/4, 1/4, 1/4 section = 10 acres

Internet Sites with Conversion Aids

www.eculine.be/metricon.htm
www.tkn.unomaha.edu/ips/tkn/hotline/helpers/metriclength.html

■ APPENDIX E—RULES FOR SUCCESSFUL FIELD TRIPS

Introduction

If you and your fellow students are lucky enough to live in an area rich in fossils or minerals and safe collecting areas, you might learn a great deal about the occurrence, preservation, and taxonomy of fossils by making your own collections and identifying your fossils. In a mineral-rich area you will have the opportunity of seeing many minerals and gaining a better understanding of the rock-forming processes.

Geologic field trips can be a lot of fun, but remember these clues when collecting fossils:

- ALWAYS have the permission of the landowner. Legal problems and insurance concerns have made many landowners and quarry operators reluctant to allow you on their property. If you were to enter a property that was posted for no admittance, you would be unprotected by insurance and subject to legal action . . . and possibly even arrest.

- Research the trip ahead of time. Check out the locality with maps, and determine how you will reach the desired location. Are the roads passable in all weather? How far away from the road is the site and is the access easy enough for all levels of hiking ability?

- Go with a friend or with a group. Besides making the trip more pleasant, in the event of an accident, it is good to have help close by.

- Never collect at the top or base of steep quarry walls. Blasting may have loosened rocks that can slip from underfoot . . . or that might fall on you.

- Take the right equipment. Carpenter's hammers are hardened for nails, not for rocks. If you pound on stone with a carpenter's hammer, you might cause steel splinters to break off the hammerhead, which could cause injury. Get appropriate rock hammers or mason's hammers designed for this task. So-called "cold-chisels" designed for cutting steel work well to split and trim rock, where wood chisels are immediately ruined.

- Wear good eye protection such as glasses or goggles. Eyes are easily damaged and hard to repair. Hard hats of the type worn by construction workers are also a very important field trip accessory.

- Plan ahead with changes of clothing, rain gear, and materials to pack your precious specimens. There is no point in collecting items of interest if they break on the trip home.

ANHEDRAL CRYSTAL: A crystal that grew in a confined space that caused it to become cramped and deformed as the growing crystal faces meet and are limited by adjacent growing crystals. The shape is more or less random. (See Chapter 1)

APHANITIC TEXTURE: Texture of an extrusive igneous rock where the magma cooled rapidly and the rock crystals are too small to distinguish or identify with the unaided eye. (See Chapter 2)

ARKOSE: Clastic rock rich in feldspar fragments. (See Chapter 3)

ASH (VOLCANIC): Tiny particles of volcanic glass produced when gasses escape during a pyroclastic eruption and produce an aerosol mist of molten magma particles. (See Chapter 2)

BAR SCALE: A distance scale that is normally located at the lower center margin of a map. The scale is arranged with a series of linear bars marked off in miles, feet, and kilometers. (See Chapter 5)

BASAL CLEAVAGE: Cleavage style present in sheet silicates (i.e., mica) where the mineral cleaves (parts) in one direction only. The cleavage is parallel to the mass of the mineral so that the cleavage fragments produced are commonly sheetlike. (See Chapter 1)

BASE LINE: Reference line used in Public Land System Survey (Township and Range Survey System). The base line is an east-west starting point for these surveys and intersects the north-south reference line called the Principle Meridian. (See Chapter 6)

BEARING (MAP): In mapping, a bearing is the position of an object or location relative to a point on a compass, that is, the compass direction between the center of the compass and a line of sight to the object being mapped. (See Chapters 5 and 7)

BENCHMARK: A survey point on a map that has been mapped with extreme accuracy to determine both the location on the earth's surface and the location's elevation above or below sea level. Benchmarks are normally the most accurately determined points on any map. (See Chapter 6)

BIOCHEMICAL SEDIMENTARY ROCK: A sedimentary rock produced from the remains of once-living organisms (such as shells, coral fragments, calcareous algae). (See Chapter 3)

BIOCLASTIC SEDIMENTARY ROCK: A sedimentary rock composed mainly from broken fragments (clasts) of once-living organisms (such as shells, coral fragments, calcareous algae). Differs from a Biochemical Sedimentary Rock in that the organisms were no longer living, having been transported, abraded, and redeposited. (See Chapter 3)

BM: Abbreviation for Benchmark (see above). (See Chapter 6)

BOMBS (VOLCANIC): A pyroclastic particle ejected as a viscous mass during a volcanic eruption. Larger than Lapilli, it can be of any size or shape, the shape being produced as it tumbles through the air. (See Chapter 2)

BOWEN'S REACTION SERIES: A chart showing the minerals that are stable at various temperatures and pressures inside the earth. The chart explains the rational classification and origins of many types of igneous rocks. (See Chapter 2)

BRECCIA (SEDIMENTARY): Breccias are composed of sharp-edged and angular rock fragments. A sedimentary breccia is commonly produced by collapse, that is, a cliff or the ceiling of a cave. (See Chapter 3)

BRECCIA (VOLCANIC): Breccias are composed of sharp-edged and angular rock fragments. A volcanic breccia is commonly produced by an explosive event in the course of a volcanic eruption, the explosion producing many rock fragments. (See Chapter 2)

BRITTLE: Materials that are brittle will not deform, bend, or change shape when stress is applied. Brittle materials will break or shatter. (See Chapter 1)

CALCARENITE: A limestone composed 50 percent or more of sand-sized calcium carbonate grains. (See Chapter 3)

CALCIRUDITE: A limestone composed 50 percent or more of calcite grains larger than 2 mm in diameter and generally well-cemented to form a dense rock; therefore essentially a calcareous gravel. (See Chapter 3)

CALCISILTITE: A limestone composed primarily of silt-sized calcite grains. (See Chapter 3)

CARBONATE ROCK: Rock composed primarily of calcite grains. (See Chapter 3)

CHEMICAL SEDIMENTARY ROCK: An organic or inorganic sedimentary rock made up entirely or nearly entirely of particles precipitated by chemical processes, that is, evaporates, chert, chemical limestones. (See Chapter 3)

CHEMICALWEATHERING: Weathering processes that chemically alter, add, or remove one or more materials in a preexisting sedimentary rock. (See Chapter 3)

CHERT: Microcrystalline quartz rock with a dull, smooth texture, conchoidal fractures, and is the rock frequently called 'flint.' (See Chapter 3)

CLASTIC SEDIMENT: Sediment composed of fragments of preexisting rocks and minerals, then transported to their place of deposition, that is, sandstone, gravel. (See Chapter 3)

CLEAVAGE: Breaking of a mineral along predetermined planes, controlled by the size and arrangement of the atomic structure of the mineral. (See Chapter 1)

COMPASS BEARING: Direction relative to north as indicated by a compass. (See Chapter 7)

COMPASS HOUSING: Structure containing and protecting the freely moveable compass needle. (See Chapter 7)

COMPASS NEEDLE: A freely moveable magnetized steel needle that can rotate in any direction, but which aligns itself to the Earth's magnetic field, indicating North and South. (See Chapter 7)

CONGLOMERATE (CONGLOMERATIC): A coarse clastic sediment composed of rounded or subangular grains that are gravel-sized or larger, set in a fine-grained matrix of sand or smaller particles, commonly cemented. (See Chapter 3)

CONTACT METAMORPHISM: Metamorphic change in a rock mass brought about by a nearby source of great heat, such as an intruding magma mass. (See Chapter 4)

CONTOUR INTERVAL: The vertical distance in feet or meters between any two adjacent contour lines. (See Chapter 6)

CONTOUR LINE: A line on a contour map connecting all points of equal altitude above or below sea level. (See Chapter 6)

CONTOUR MAP: A map demonstrating the three dimensional shape of the land as determined by contour lines. (See Chapter 6)

COQUINA: Rock composed mainly of shells and shell fragments. Coquina is also of low density and contains numerous pores and cavities, distinguishing it from a calcirudite. (See Chapter 3)

COUNTRY ROCK: Term describing the general mass of rock layers or materials present and surrounding some other unit of interest, that is, the preexisting rock masses, which were intruded by a rising mass of molten magma, would be described as the country rock. (See Chapter 2)

CRYSTAL SETTLING: Process where dense crystals appearing in a mass of molten magma may settle out of the mass, forming layers across the bottom of the magma chamber, thereby depleting the magma of that mineral, changing the nature of the resulting rock when the rest of the magma cools. **EXAMPLE:** Olivine crystals form early and may settle out, leaving the magma depleted in some mafic constituents. (See Chapter 2)

CUBE: Shape with six faces, with a 90 degree angle between each pair of adjacent faces. (See Chapter 1)

DEGREE (OF ARC): Circles can be divided into 360 equal segments. Each segment is one degree. A degree consists of 60 minutes of arc. (See Chapter 5)

DENSITY: Also referred to as specific gravity or mass of a material. Density is the ratio of weight per unit of volume. Density measurements use water as a standard; therefore, one unit of water at room temperature and sea level has a density of one, and all other materials are either lighter, equal to, or heavier than this standard. Density is usually stated in grams per cubic centimeter. (See Chapter 1)

DETRITAL ROCK (DETRITUS): Detritus is a general term for loose rock and other clastic material derived by erosion or other mechanical means from preexisting rock masses, then moved to some new place of deposition. Detrital rocks are therefore composed of at least 50 percent detritus. (See Chapter 3)

DOLOSTONE: A rock unit primarily composed of the mineral dolomite (calcium magnesium carbonate). (See Chapter 3)

DOUBLE REFRACTION: Seen in some minerals, such as optical calcite, where two images are seen when an object is viewed through the mineral mass. The double images are generated because light travels through the mass in two pathways determined by the bending of light rays within the mineral, causing one of the optical pathways to travel further and arrive later than the other. (See Chapter 1)

EFFERVESCENCE: Term describing the bubbling seen when hydrochloric acid is applied to some rocks, such as calcite or dolomite. Bubbles are produced by the release of carbon dioxide as the rock is dissolved by the acid. (See Chapter 1)

ELASTIC: A mass that is elastic will deform or bend when a stress is applied to it, but will return to its original shape when that stress is removed, that is, the behavior of a piece of rubber when bent or stretched. (See Chapter 1)

EUHEDRAL CRYSTAL: Crystal formed in a space sufficient for the crystal to grow without interference, thereby allowing all the crystal faces to develop their optimum shape. Generally seen as a "perfect" crystal shape, as opposed to subhedral or anhedral crystals. (See Chapter 1)

EXTRUSIVE IGNEOUS ROCKS: Fine-grained crystalline igneous rocks. The small grain size is generated by the relatively rapid cooling as the mass of magma is extruded onto the surface of the earth, such as during a volcanic eruption. Grain size is a function of the cooling rate, and extrusive rocks cool quickly. (See Chapter 2)

FAULT ZONE METAMORPHISM: Metamorphic rock changes produced along the actual fault plane due to the pressures and temperatures generated when the rocks on either side of the fault plane moved. (See Chapter 4)

FELSIC: Term used to describe those igneous rocks high in quartz and potassium feldspars, hence rocks that are generally granite like. Felsic rocks are generally light in color and relatively low in density, in contrast to intermediate or mafic rocks that are darker and more dense. (See Chapter 2)

FERROMAGNESIAN: Said of minerals rich in iron and magnesium. Ferromagnesian minerals and rocks are dark in color and relatively high in density. (See Chapter 2)

FLEXIBLE: Flexible materials can be bent or deformed when a stress is applied, but when that stress is removed, flexible materials remain deformed or bent, unlike the condition seen in elastic materials (see above). (See Chapter 1)

FISSILE: Said of fine-grained sedimentary rocks, such as shales, siltstones, and similar materials where the bedding is very regular and the rock masses tend to split apart in sheetlike masses, much like the pages in a book. (See Chapter 3)

FOLIATED METAMORPHIC ROCK: Foliated metamorphic rocks are composed of sheetlike minerals or bands of minerals. These rocks tend to split into layered masses (as in slates, phyllites, or schists) or show strong color banding (such as seen in gneiss). (See Chapter 4)

FRACTURE: Breakage in a mineral that produces irregular and random shapes, the opposite of cleavage. (See Chapter 1)

GEOGRAPHIC NORTH OR SOUTH POLE: These are the points determined by the rotational axis of the planet, and are different than the locations of the north or south magnetic poles. All maps are created relative to the north and south geographic poles. (See Chapter 5)

GNEISS: Metamorphic rock characterized by parallel to sub-parallel color banding, where the bands are composed of different colored minerals. (See Chapter 4)

GRADIENT: Term used to describe the sloping surface down which a stream is flowing. Described in vertical distance dropped over some horizontal distance traveled, for example, a stream gradient of two feet per mile. (See Chapter 6)

GRAYWACKE (WACKE): Generally described as a 'dirty' sandstone. Sandstone rich in silt and clay sized material, often dark in color. (See Chapter 3)

GREENWICH MERIDIAN: The north-south line of longitude running through Greenwich Palace (essentially part of modern London, England) which defines the point of zero longitude. All other lines of longitude are described as east or west longitudes relative to the Greenwich Meridian. (See Chapter 5)

GRID NORTH: Term used for determining location of a spot on the globe. Similar in concept to longitude and latitude. Most commonly used by military personnel and less commonly by civilian surveyors. (See Chapter 5)

GROUNDMASS: The general matrix material within a sedimentary rock or the finer-grained materials between the coarse crystals within a porphyry's phenocrysts. (See Chapter 2)

HACHURE: Mapmakers' marks to suggest a closed depression on a contour map. Hachures are short dashed marks at right angles to the contour lines and pointing into the deeper parts of the depression. (See Chapter 6)

HEXAGON (HEXAGONAL): A hexagon is a shape commonly found in some minerals where the cross section of the crystals has six faces, with angles between each face being 60 degrees. Some crystals, such as quartz, may also show a pointed or pyramidal tip with six faces. (See Chapter 1)

HYALINE (GLASSY) TEXTURE: Rock or mineral texture where the fractured or broken surfaces resemble those of broken glass. (See Chapter 1)

HYDROTHERMAL ALTERATION: Changes in a rock produced by the effects of hot water, with or without other chemicals present. Common around the edges of an igneous intrusion. (See Chapter 4)

IGNEOUS ROCK: Igneous rocks began their existence as a mass of molten materials that crystallized into one of a number of end products, such as granite, basalt, etc. (See Chapter 2)

INDEX CONTOUR: In drawing contour lines of a map, every fifth line is normally drawn with a heavier line and is marked with the elevation of that line. This heavier line is the index contour. (See Chapter 6)

INORGANIC: Said of materials or chemicals that are normally produced by processes that do NOT involve living plants or animals. (See Chapter 1)

INTERMEDIATE IGNEOUS ROCK: Igneous rock that contains little or no quartz, little or no potassium feldspar, significant plagioclase feldspar, and significant quantities of ferromagnesian minerals. Normally these rocks are darker in color and are more dense than felsic igneous rocks. Examples include diorite and andesite. (See Chapter 2)

INTERNATIONAL DATE LINE: Imaginary line drawn on the globe at the point where the east and west longitude lines meet (each at 180 degrees). Used by convention to determine where the international dates and times change. (See Chapter 5)

INTERPOLATION: Process whereby two or more points on a map are known with great precision, and those points are used to estimate the location of other points. For example, in making contour maps, if the spot on the map representing an elevation of 150 feet and a spot representing 250 feet are both known, the spot representing 200 feet can be precisely estimated by locating a spot halfway between the two known locations. (See Chapter 6)

INTRUSIVE IGNEOUS ROCKS: Rocks produced by masses of liquid magma that has cooled within the mass of the earth. These rocks are insulated by the surrounding rocks and therefore cool slowly, allowing the crystals to grow to relatively large size, thereby producing a coarse-grained texture. (See Chapter 2)

JOINT: Place within the earth where a fracture occurs WITHOUT movement (offset). (See Chapter 15)

LAPILLI: A pyroclastic rock fragment produced during the explosive phase of a volcanic eruption. Lapilli are 1–64 millimeters in diameter, but are commonly the general size and shape or garden peas. (See Chapter 2)

LATITUDE: Imaginary line drawn on the globe parallel to the equator. Lines of latitude are measured north 90 degrees to the north geographic pole and south 90 degrees to the south geographic pole. Lines of latitude cross lines of longitude at right angles. (See Chapter 5)

LAVA: Term applied to molten magma when it spills out onto the surface of the earth, such as the flows that emanate from a volcanic eruption. (See Chapter 2)

LOCAL RELIEF: The vertical distance between two points of elevation within an area being studied, that is, the difference in altitudes between the top of a local hill and the bottom of an adjacent valley expressed in meters or feet. (See Chapter 6)

LONGITUDE: Imaginary lines drawn on the globe from the north geographic pole to the south geographic pole, crossing the equator and other lines of latitude at right angles. They are numbered starting with zero degrees at the Prime Meridian (Greenwich Meridian) and increase to 180 degrees west or 180 degrees east of the Prime Meridian, meeting on the opposite side of the globe at the International Date Line. (See Chapter 5)

LUSTER: The manner in which light is reflected from a broken surface of a rock or mineral, that is, metallic, resinous, hyaline, adamantine, etc. (See Chapter 1)

MAFIC IGNEOUS ROCK: An igneous rock rich in iron and magnesium, generally dark to very dark in color and high in density relative to Felsic and intermediate rocks. (See Chapter 2)

MAGMA: Generic term applied to a mass of igneous rock while still molten. (See Chapter 2)

MAGMA MIXING: Condition known where two or more magma chambers filled with fluid igneous rocks become connected in the subsurface and one flows into the other, changing the character and appearance of the resulting solid rock. Essentially one magma body flows into another and a mixture is formed. (See Chapter 2)

MAGNETIC DECLINATION: The difference in degrees seen on a magnetic compass between the location of true geographic north and magnetic north. (See Chapter 5)

MAGNETIC NORTH OR SOUTH POLE: The location of the point on the globe indicated by the north end of a compass needle is the North Magnetic Pole. The north geographic pole is currently located in the Arctic Sea north of the Canadian Arctic Islands and near the northwestern tip of Greenland, approximately 82 degrees North latitude and 113 degrees West longitude. The South Magnetic Pole is 180 degrees away. (See Chapter 5)

MALLEABLE: Malleable materials are able to be plastically deformed when compressive stress is applied, changing shape as in metals such as hammering produces in gold, silver, iron, etc. (See Chapter 1)

MECHANICAL WEATHERING: Processes whereby a rock material is broken down into smaller particles without changing the chemical nature of the rock. Examples include frost wedging, exfoliation, thermal shock, etc. (See Chapter 3)

METAMORPHIC GRADE: A measure of the degree of alteration in a metamorphic rock. An expression of how intense the heat, pressure, and chemical action has been in the generation of a particular type of metamorphic rock. Low-grade rocks are only slightly altered, while high-grade metamorphic rocks are extensively altered. (See Chapter 4)

METAMORPHIC ROCK: Rock of any type that has been altered by the application of great heat, pressure, and chemically active fluids at depth within the crust of the earth so that new minerals and textures are generated. (See Chapter 4)

METES AND BOUNDS MAPPING: Type of surveying techniques involving the locations of well-known local landmarks, specific trees, buildings, rocks, etc. Mainly found today in the states formed from the original thirteen colonies, Texas, and a few other sites in North America. (See Chapter 5)

MICRITE: Rock produced from chemically precipitated calcite mud. Micrites are generally dense and very fine grained, such that no evident visible texture can be determined without magnification. (See Chapter 3)

MINERAL: Any element or compound that: has a definite chemical formula, has a definite internal physical structure, is crystalline, is inorganic, and is naturally occurring (that is, not man-made). (See Chapter 1)

MINUTE (OF ARC): A degree consists of 60 minutes of arc. (See Chapter 5)

MOHS HARDNESS SCALE: A measure of relative hardness for all minerals. There are ten steps from talc (softest) to diamond (hardest). Each step is ten times harder than the one below. (See Chapter 1)

MUDSTONE: Very fine grained siliciclastic rock, composed of clay or clay-sized grains. Differs from a shale by breaking in irregular chunks and masses. Not fissile. (See Chapter 3)

NONFOLIATED METAMORPHIC ROCK: Metamorphic rock generally lacking sheetlike minerals or color banding, being mostly granular or fine grained, and of simple composition. The opposite of foliated metamorphic rocks. (See Chapter 4)

OCTAHEDRON (OCTAHEDRAL): An octahedron is a crystalline form bounded by eight faces and produces the shape of a pair of pyramids cemented together base to base. Commonly seen in cleaved examples of fluorite. (See Chapter 1)

OOLITE (OOLITIC): Oolites are small rounded to oval grains of laminated calcite produced in the surf or other wave-disturbed zones within high evaporation rate tropical seaways. Oolitic rocks generally appear to be composed of small white or tan spheres. (See Chapter 3)

ORGANIC: Said of materials or chemicals that are normally produced by processes that involve living plants or animals. (See Chapter 3)

PACE AND COMPASS MAPPING: Simple surveying method involving a hand-held compass and a known walking pace (stride) length to determine and measure a series of azimuths and boundaries of an area of interest. Useful in beginning a survey, but subject to many inaccuracies due to grade, surface plant cover, and topography. (See Chapter 7)

PACE LENGTH (PL): Used in pace and compass mapping. Walking in a normal manner across a flat surface, a person can measure the distance between two successive foot positions (either left and right foot falls, or two strides giving the measurement between two successive right or left foot falls). This distance in inches or meters is the pace length. (See Chapter 7)

PARTIAL MELTING: Condition where a preexisting rock is subjected to great heat and pressure such that some of the components can be melted and removed. In such a case involving a granite, a point of temperature and pressure would occur where the quartz and potassium feldspars might melt and be drained away, leaving behind the remainder of the higher melting point minerals to form a new rock type. (See Chapter 2)

PEGMATITE (PEGMATITIC TEXTURE): Rock generated when a magma cools very slowly, allowing the crystals to grow to exceptionally large sizes. The mineral grains are generally more than one centimeter in diameter, although some authors suggest two centimeters or more in diameter. (See Chapter 2)

PHANERITIC TEXTURE: Texture found in igneous rocks where the cooling rate was slow enough to allow the crystals to grow sufficiently large to be easily seen and recognized with the unaided eye. (See Chapter 2)

PHENOCRYST: Condition where one or more of the mineral types present has grown significantly larger than the remaining portion of a rock, thereby being very visible and evident. Generally used for the large crystals seen in a porphyry. (See Chapter 2)

PHYLLITE: Moderately fine-grained foliated metamorphic rock where the mica grains are large enough to reflect like, giving the rock a silky sheen. (See Chapter 4)

PLANOMETRIC MAP: The common type of map used by most persons in an atlas or as the "gas station" map used by travelers. Planometric maps show the precise location of places and features such as rivers and coastlines, but lack information about the shape or topography of the landscape. (See Chapter 5)

PLS SURVEY: Public Land Survey System is a synonym for the mapping system otherwise known as Township and Range Mapping. (See Chapter 6)

PORPHYROBLAST: In metamorphic rocks of schist-grade, some mineral crystals (such as garnet, staurolite, etc) may grow much larger than the other mineral grains and become porphyroblasts that are very visible and evident. (See Chapter 4)

PORPHYRY (PORPHYRITIC): A rock that formed with two very different cooling rates. One cooling rate was slower and occurred when the magma was deeply buried and generated larger crystals, while in a later event the magma became extrusive and the remaining fluid cooled quickly, producing a fine-grained rock matrix between the earlier large crystals. The resulting rock is a porphyry and is generally named for the matrix rock type, that is, an andesite porphyry. (See Chapter 2)

PRECURSOR ROCK: In studies of metamorphic rocks, the precursor rock is the material that was altered during metamorphism. (See Chapter 4)

PRIME MERIDIAN: The north-south line of longitude running through Greenwich Palace (essentially part of modern London, England), which defines the point of zero longitude. All other lines of longitude are described as east or west longitude relative to the Greenwich Meridian. (**CAUTION:** DO NOT MISTAKE THIS FOR THE PRINCIPAL MERIDIAN). (See Chapter 5)

PRINCIPAL MERIDIAN: The starting point for PLSS or Township and Range Surveys depends on two surveyed lines: the base line, an east-west reference line, and a north-south reference line that crosses the base line at a definite point. The north-south reference line is the Principal Meridian. (**CAUTION:** DO NOT MISTAKE THIS FOR THE PRIME MERIDIAN). (See Chapter 6)

PROVENANCE: Termed used for the original source location of a sedimentary or other detrital material, that is, much of the western Great Plains of the United States had its provenance in the front range of the Rocky Mountains. (See Chapter 4)

PUBLIC LAND SURVEY SYSTEM: A synonym for both the PLS System and the mapping system otherwise known as Township and Range Mapping (see below). (See Chapter 6)

PUMICE: A generally felsic pyroclastic rock that is highly porous, often so low in density that a sample will float. (See Chapter 2)

PYROCLASTIC ROCK: Rock material generated during the explosive eruptive phase of a volcanic event. The particle sizes produced are given special names, that is, ash, Lapilli, bombs, etc. (See Chapter 2)

QUADRANGLE MAP: Generic name given to maps of an area bounded by longitude and latitude. The north-south and east-west dimensions are not the same, being determined by the location relative to the equator. The map sizes are given in minutes or degrees, that is, 7.5 minute, 15 minute, etc. (See Chapter 5)

QUADRANT (MAP): In reference to a quadrangle map, referring to an arc of 90 degrees. Therefore, a map can be divided into four parts by a north-south and an east-west line, producing in the upper right corner the northeast quadrant, in the lower right-hand corner is the south-east quadrant, in the lower left-hand corner is the south-west quadrant and in the upper left corner is the north-west quadrant. (See Chapter 5)

RANGE LINE: In PLS Survey or Township and Range Surveys, the range lines are those surveyed edges of townships parallel to the principal meridian. (See Chapter 6)

RATIO (MAP) SCALE: Used on quadrangle maps to determine the scale on the map relative to the ground. The scale uses a ratio of one unit on the map being equal to some stated number of those same units on the ground. For example, a scale states as 1:24,000 means that one unit (centimeter or inch) on the map is equal to 24,000 centimeters or inches on the ground. (See Chapter 5)

REGIONAL METAMORPHISM: Large-scale metamorphic event generated over a broad geographic area, generally the result of mountain building tectonic events. (See Chapter 4)

RELIEF: The total difference in altitude between the highest and lowest points in an area being studied, expressed in feet or meters above or below sea level. Also used generally as an expression of the shape of the land. **EXAMPLE:** This area has a high relief or low relief, meaning that the one has high hills or mountains while the other is flat or gently rolling. (See Chapter 6)

RHOMBOHEDRON (RHOMBOHEDRAL): Three-dimensional shape with six sides generally resembling a leaning rectangle. Seen in some common minerals such as calcite and dolomite. (See Chapter 1)

RULE OF V'S (CONTOUR MAPS): In preparing or reading contour maps, one should note the condition where a stream crosses the contour lines. The contour line ALWAYS forms a "V" shape towards the upstream portion of the stream. (See Chapter 6)

SCHIST: Medium-grained metamorphic foliated rock with scaly texture, and often with porphyroblasts. (See Chapter 4)

SCORIA: A generally mafic (reddish to black) rock that is highly porous, but normally will not float. (See Chapter 2)

SECOND (OF ARC): A degree consists of 60 minutes of arc, and each minute of arc is further divided into 60 seconds. (See Chapter 5)

SECTION: Term used in PLSS or Township and Range Mapping for a piece of surveyed land one mile on a side, therefore containing one square mile, or 640 acres. There are 36 sections in a Township. (See Chapter 6)

SEDIMENTARY ROCK: A rock unit resulting from the accumulation and consolidation of sedimentary materials, such as clastic grains, chemical grains, detritus, etc. The processes can involve wind, water, gravity, chemical activity, or ice. (See Chapter 3)

SHALE: Very fine grained siliciclastic rock, composed of clay or clay-sized grains. Differs from a mudstone by being fissile. (See Chapter 3)

SILICICLASTIC SEDIMENT: A sediment consisting almost entirely of clastic particles, such as clays, siltstones, sandstones, gravels, sedimentary breccias, etc., where the materials are almost exclusively silicon based (quartz-rich commonly). (See Chapter 3)

SILTSTONE: Very fine grained siliciclastic rock, composed of silt-sized grains. Generally not fissile. (See Chapter 3)

SLATE: Very fine-grained metamorphic foliated rock with no apparent texture, splits easily into sheets and is found in many colors. (See Chapter 4)

SPECIFIC GRAVITY: Also referred to as the density or mass of a material. Density is the ratio of weight per unit of volume. Density measurements use water as a standard; therefore, one unit of water at room temperature and sea level has a density of one, and all other materials are either lighter, equal to, or heavier than this standard. Density is usually stated in grams per cubic centimeter. (See Chapter 1)

SPOT ELEVATION: Local surveyed point on a map, generally indicated by an "x" accompanying an elevation number. These are thought to be accurate, but have not been determined as carefully as the benchmark elevations. (See Chapter 6)

STREAK: Refers to the color of powdered mineral left on a streak plate (piece of unglazed porcelain) as an aid to identifying an unknown mineral. (See Chapter 1)

STRIATIONS: Series of shallow parallel depressions or grooves on the cleavage faces of some minerals, especially plagioclase feldspars. (See Chapter 1)

SUBHEDRAL CRYSTAL: Crystal that was not able to grow freely, but grew against other crystals sufficiently to deform the growth pattern and produce less than perfect shapes. (See Chapter 1)

TEXTURE (ROCK): The physical appearance, grain size, and arrangement of particles. In classification of igneous rocks, samples are defined as having phaneritic, aphanitic, pegmatitic, porphyritic, glassy, frothy, vesicular, and pyroclastic textures. (See Chapter 2)

TIER LINE: Line parallel to the base line and forming the north or south side of a township, as used in the PLS Survey (Township and Range Survey). Synonymous with Township Line. (See Chapter 6)

TOPOGRAPHIC MAP: A map demonstrating the three-dimensional shape of the land as determined by contour lines. Synonymous with Contour Map. (See Chapter 6)

TOPOGRAPHIC PROFILE: A diagram or drawing representing the profile across some area of interest. Determined by drawing a line of traverse and locating the contour intervals along the line, transferring these points to a grid, calculating a percentage of relief exaggeration, and plotting a line that represents the profile or relief of the land being studied. (See Chapter 6)

TOTAL RELIEF: As opposed to local relief, total relief is the difference in elevation between the highest location in a region and the lowest site. (See Chapter 6)

TOWNSHIP: In PLS Surveys and Township and Range Mapping, one of the major area subdivisions is the township, which is an area composed of thirty-six sections. Each section contains one square mile, therefore a township contains thirty-six square miles and is six miles (or sections) wide and high. (See Chapter 6)

TOWNSHIP AND RANGE MAPPING: A system of mapping where the starting reference points are an east-west line called a base line and an intersecting north-south line called a principal meridian. At six-mile intervals away from each of these lines, range lines and tier (or township lines) are established, generating townships. Townships are further subdivided into sections. (See Chapter 6)

TOWNSHIP LINE: Line parallel to the base line and forming the north or south side of a township, as used in the PLS Survey (Township and Range Survey). Synonymous with Tier Line. (See Chapter 6)

TUFF (VOLCANIC): One of the materials classified as a pyroclastic rock. An aerosol mist of volcanic glass particles is produced as gasses are released at high pressure through a mass of molten magma in the throat of a volcano. This generates volcanic ash, which can accumulate, and when compacted and cemented to form a dense rock, produces volcanic tuff. (See Chapter 2)

ULTRAMAFIC IGNEOUS ROCK: Rock consisting entirely or almost entirely of ultramafic minerals. Such rocks are commonly olivine rich, dark in color, and especially dense when compared to other igneous rocks. (See Chapter 2)

VERBAL (MAP) SCALE: Method of determining and stating the scale of a map, by translating it into a verbal statement, that is, "one inch equals one mile." (See Chapter 5)

VERTICAL EXAGGERATION: When preparing topographic profiles and other large-scale regional studies, it will often be found that the areas studied have too little relief to be useful graphically. A more easily perceived profile can be prepared by deliberately adding some percentage of exaggeration to the relief measurements. (See Chapter 6)

VESICULAR: The condition of an igneous rock filled with holes (vesicles) representing gas bubbles trapped there when the rock cooled. Two examples of vesicular rocks are scoria and pumice. (See Chapter 2)

WENTWORTH SIZE SCALE: Standard scale of grain sizes linked with names used by engineers, builders and geologists. (See Chapter 3)